The National Computing Centre develops techniques, provides services, offers aids and supplies information to encourage the more effective use of Information Technology. The Centre co-operates with members and other organisations, including government bodies, to develop the use of computers and communications facilities. It provides advice, training and consultancy; evaluates software methods and tools; promotes standards and codes of practice; and publishes books.

Any interested company, organisation or individual can benefit from the work of the Centre – by exploring its products and services; or in particular by subscribing as a member. Throughout the country, members can participate in working parties, study groups and discussions; and can influence NCC policy.

For more information, contact the Centre at Oxford Road, Manchester M1 7ED (061-228 6333), or at one of the regional offices: London (01-353 4875), Bristol (0272-277 077), Birmingham (021-236 6283), Glasgow (041-204 1101) or Belfast (0232-665 997).

Do You Want to Write?

Could you write a book on an aspect of Information Technology? Have you already prepared a typescript? Why not send us your ideas, your 'embryo' text or your completed work? We are a prestigious publishing house with an international reputation. We have the funds and the expertise to support your writing ambitions in the most effective way.

Contact: Geoff Simons, Publications Division, The National Computing Centre Ltd, Oxford Road, Manchester M1 7ED.

The Technical Documentation Handbook

A J Marlow

PUBLISHED BY NCC PUBLICATIONS

British Library Cataloguing in Publication Data

Marlow, A. J.
The technical documentation handbook
1. Technical Writing — Manuals
I. Title
808'.0666921

ISBN 0-85012-704-1

© THE NATIONAL COMPUTING CENTRE LIMITED, 1988

All rights reserved. No part of this publication may be reproduced, stored in a retrieval system, or transmitted, in any form or by any means, without the prior permission of the National Computing Centre.

First published in 1988 by:

NCC Publications, The National Computing Centre Limited, Oxford Road, Manchester M1 7ED, England.

Typeset in 10pt Maximal.
Printed by Hobbs the Printers of Southampton

ISBN 0-85012-704-1

Acknowledgements

The author wishes to thank all those who have helped in the production of this book. While it is not practical to acknowledge each individual, thanks are due to those organisations who have provided reference material for the appendices and members of Pegasus Software Limited who have provided the means which enabled this project to be realised.

Special mention should be made, however, of some key people: Johnnie Johnson (Pegasus) for putting me in this profession, Clive Booth (Pegasus) for keeping me there, and Geoff Simons (NCC) for accepting the proposal and seeing it through.

Thanks are also due to Digitext for allowing me to quote from their National Documentation Survey for the computer industry, and Phoenix Technical Publications.

Finally, a very special thanks to Brenda Durndell for her invaluable assistance throughout the project.

Preface

This book is aimed at anyone involved in the publication cycle of technical documentation in the computer industry. Whether you are an author, publisher, publications manager, or the general dogsbody who just happens to have been landed with the task of producing some documentation (or any aspect of it), or perhaps your involvement is a single string to your bow of responsibilities, you'll find something to read in this volume which may help you. Just how much help you get from the reading will depend on a number of factors. Your own experience for one. Your freedom of control over the work you do for another.

The computer industry is without doubt becoming increasingly reliant upon well produced and clear documentation to accompany its expanding range of technology products and services. Your own experience will show you just how much variety there exists in terms of quality, presentation, format and style of documentation across the industry; these range from the very worst examples of written material to some commendable, easy-to-understand manuals and brochures.

In the inaugural issue (Autumn 1987) of *The Stag* — the official newsletter of the Science, Technology and Specialist Group of the Society of Authors, the group's chairman Brian J Ford commented '.. the most advanced publications in that field (computer manuals) are crammed with facts — yet frequently written in the most obscure, muddle-headed and illiterate manner imaginable. It is (in) this field that the worst writing of all can be found...'.

Harsh words for the technical authors of the industry, but the responsibilities do not lie solely at their feet; neither is the writing alone to blame for a poor reception of documentation in the market place. A document can be well written, but through poor production will do anything but invite the user to read it. After all, my copy of *The Stag*, which on the back page proudly informed me how the edition was produced using such technology as a BBC Master Compact Disk and View Word Processor, transferred to an Apple Macintosh and typeset through Pagemaker software and an Apple Laserwriter, had a poor appearance on account of the fact that it had been photocopied, and suffered considerable degradation of quality in the process.

It is worth mentioning that the views expressed in this publication are solely that of the author and do not necessarily reflect the opinion of the publishers. Some of the more subjective material within this volume you may disagree with, but if it helps you believe that you've got it right, then all's well and good. Moreover, the book serves as a reference guide as well as a guidebook. The appendices contain a few details of organisations in the computer industry which have something to offer in the field of technical documentation — whether it is tools for the trade or services to do the job for you, train you or your staff, and so on. If you work in such an organisation and find in horror that your company is not included in the appropriate appendix, then I apologise. The information is by no means comprehensive, but is based upon the contacts the author has during the experience of writing and controlling technical documentation. Rather than be an extensive listing, the appendices serve to start one off in the right direction towards getting help in some sphere of the profession, and to this extent, necessarily exclude many possible sources that may be more local to the reader.

I have had the pleasure of listening to many authors and publications managers in the business about their specific problems in this field of work. It has been possible to identify some common ground between many of them, and it has become apparent that most of their difficulties arise from their relationship with other departments in their own organisations, and getting the appropriate recognition for their work, compared to, say, development personnel, technicians and the like.

Often, those who work in the field of technical documentation production sit on a bridge between the marketing forces of a company and the technical departments, but don't wholly fit into either. This should not be a problem, but is perceived as being one. I have often felt that, because publications managers or technical writers feel themselves to be in a unique position within an organisation, with unique problems to overcome, that due respect should be given to them by their colleagues in a similarly unique way.

I have included two chapters in the book, one for authors and author/managers and one for publications managers, about their roles in the organisation for which they work and how they can help themselves to be recognised for what they are and do, yet at the same time realising their own responsibilities to the company — a point often overlooked, since human nature seems to favour our being interested in our own self-importance. Those involved in publications of technical documentation have a responsibility to improve performance, increase efficiency, reduce costs and maintain budgets, meet deadlines and issue schedules, just like any other department in a commercial organisation, and some parts of this book have touched on such issues.

You should understand that the book does not aim to teach technical writing per se. The sections on form and style, for example, discuss different approaches but do not aim to teach or preach any particular approach. This is largely because it is especially highly subjective, and in any case there are plenty of other sources of tuition in this field (also subjective) in written form and through the training courses mentioned in the appendices. Similarly, the chapters discussing technology do not recommend any particular approach to a solution. Rather, some of the more widely used systems are appraised for their individual merits, and disadvantages are given equal weight to their advantages.

This is particularly so regarding desktop publishing technology. It is still a technology in its infancy on the microcomputer but is set to become a very significant, even commonplace, resource in the publications office, but you would do well to heed the warnings in Chapter 7 before considering that typesetting with desktop publishing is a natural progression from word processing.

A section on printers and publishing is intended as a guide for those who do not get involved in print technology on a day-to-day basis. I am still surprised to learn from among my many aquaintances in this field, that those whose job it is to publish technical documentation still do not understand what happens to it when it reaches the hands of the printers. This lack of knowledge can be a positive drawback when it comes to good planning, especially where costs have to be considered carefully (and I can't think of any private organisation where they are not) and where a particular job can be suited or matched to a particular print process. How many of you can honestly admit to including in your design schedule the implications of using different paper types and qualities for a publication before picking your typeface, use of colour and effect on packaging costs?

Finally, I should perhaps offer one further note to the reader. The chapters are arranged in an attempt to achieve an underlying unity which may have been difficult to produce with a different structure. While the reader may find a surprisingly wide variety to the order in which the various chapters may be read, depending upon your specific interests, you will find that each chapter is divided into sections for easy reference.

A comprehensive index should help you locate specific topics or items of interest, and although a Glossary is included, I have tried to identify any jargon which I may have inadvertently used and offered an explanation where it occurs.

A J Marlow MSc
Kettering, 1988

Contents

Page

Acknowledgements

Preface

PART 1 — THE AUTHOR

1 The Author's Role 11

    Subject Knowledge 12
    Reader Knowledge 13
    English Skills 14
    Attention to Detail 16
    Communication Skills 16
    Less Appreciated Qualities 17
    A Few Questions to Ask Yourself 18
    To Conclude 19

2 Writing Sources 23

    Using Outside Sources 24
    Communicating with Internal Sources 29
    Assessing the Competition 31
    In Conclusion 31

3 Format and Style 33

    General Use of English 34
    Points to Consider 36
    House Styles 40
    Including Illustrated Concepts 43

	Structure of the Text	43
	In Conclusion	44
4	**Design and Layout**	**47**
	Introduction	47
	Layouts	48
	Design	48
	Planning the Type Area on the Page	56
	Chapter Beginnings and Headings	61
	Footnotes and Running Heads	61
	Some Fundamentals of Type	65
	Concluding Design and Layout	72

PART 2 — THE MEANS

5	**Illustration**	**75**
	Scanners	76
	Graphics/Drawing Software	78
	CAD/CAM Systems	80
	Line Copy/Drawings	80
	Halftones	81
	Combining Line with Halftones	83
	Retouching	83
	Screen Illustrations for Software Documentation	85
	Summary	88
6	**Word Processing**	**91**
	Word Processor Types	92
	WYSIWYG	92
	Printers	93
	Using Word Processors for Phototypesetting	96
	Layout and Word Processors	96
	Word Processors for Indexing	98
	Using Electronic Proof Readers	99
	Drafting Copies on the WP	103
7	**Desktop Publishing**	**107**
	Introduction	107
	Who Are DTP Users?	110

	Identifying Your Needs	111
	Making Good Use of DTP	113
	How a Laser Printer Works	114
	Cautions for DTP Users	116
8	**Phototypesetting**	121
	Introduction	121
	Why Use Phototypesetting?	123
	How Phototypesetters Work	124

PART 3 — PUBLICATIONS MANAGEMENT

9	**The Publications Manager**	127
	Planning the Documentation	128
	Recruiting and Developing Authors	129
	Evaluating Documentation	131
	Print Buying	133
	Justifying In-house Production	138
	Costing and Scheduling	139
	Progress Chasing	139
	Stock Control	139
10	**Costing**	141
	Budgets	142
	Costing Resource Time	143
	Time Sheets	146
	Material/Overhead Cost Sheets	146
	External Costs	147
	Cost Per Page	147
11	**Scheduling**	151
	Building a Schedule	151
	General Design Concept	152
	Authoring First Draft	155
	Word Processing or Typing	155
	Proofing and Amending	155
	Illustration	155
	Editing	156
	Artwork Preparation	156

| | Printing and Finishing | 156 |
| | Using Electronic Scheduling Systems | 159 |

12	**Printers and Printing**	161
	Letterpress	162
	Gravure Printing	163
	Offset Printing (Lithography)	166
	Considering Colour Printing	168

13	**Presentation and Packaging**	169
	Format	169
	Colour	170
	Illustrations	170
	Paper Quality and Method of Printing	170
	Binding	171
	In Conclusion	172

PART 4 — LOOKING AHEAD

14	**Revisions and Updates**	173
	Design for Change	173
	Methods of Presenting Updates/Revisions	174

| 15 | **On-line Documentation** | 179 |

| 16 | **Development** | 183 |

Appendix

A	Contractors	187
B	Author Recruitment	191
C	Author Training	195
D	Copy Editing and Proof Correction	201
E	DTP Packages	205
F	Interfacing Typesetters	229
G	Organisations	237
H	Further Reading	243
I	Glossary	245

Index 253

1 The Author's Role

The view of the role of a technical writer in the computer industry has changed dramatically over just the past few years. Such a previously obscure role has become a recognised profession of equal standing to programmer/analysts. This may not be a widespread recognition, but it is now growing, and the acceptance of the importance of the author in the industry means it is no longer the stigma that it once was.

However, the position or status of the author may vary widely from one company to another, and this has certainly been borne out through conversations with the many individuals in this profession whom I have had the opportunity and pleasure of meeting.

For some, the difficulty still remains of securing for themselves, and their work, the same degree of respect and commitment as that received by other departments within their companies. In this situation lies the source of many of their problems regarding the degree of success or failure in being able to make a significant enough contribution to producing good quality documentation.

It may not be apparent to that individual's superiors precisely what the role of the technical writer is. This is, I believe, due to the fact that there are few senior managers in organisations who have had any experience in this field of work. In many cases, the author reports to either a technical department superior, or to a marketing executive. In either case, their own understanding of the requirements of the technical writer may be limited, and so the writer has to accept that this may cause difficulty when it comes to relating to their way of thinking. For marketing executives, the role of the writer is often seen primarily as a technical skill, accompanied by some flair for clear analytical thinking and good English skills.

This may be partly true, but there is more to the job than that, as any experienced author knows. In a small organisation, the role of the author often extends to having responsibility for the production requirements of the documentation itself, as well as print buying, scheduling and other productivity controls.

In such cases, the technical writer rarely gets a fair deal. But this need not be so if the individual learns a little of how to manage the responsibilities which are his or hers, and properly project requirements and results to the management.

So what are the main factors that determine the role of a technical writer or author? We will now examine a few possibilities.

SUBJECT KNOWLEDGE

It is generally accepted that the writer of technical documentation should have at least some knowledge of the subject about which he or she will be writing. The degree of knowledge required will depend upon the amount of input to documentation that may be provided by other parties. There are those who believe that the author must know, inside out, the subject about which he or she is writing if the documentation is to be at all meaningful and comprehensive.

However, good documentation can be provided by an author who has only a general grasp of the subject yet who is able to extract accurate details from those sources where subject knowledge abounds, and can then present this information in an informative way. In this latter case, the individuals who are the source of technical detail may then have to be involved in the proofing cycle of a documentation project to clarify that the author has not misinterpreted the facts nor, by lack of understanding, misrepresented the material in the way in which the facts have been assembled and projected through the writing. But this is not a disadvantage.

THE AUTHOR'S ROLE

If an author is to be flexible enough to be capable of writing more than simply specialist literature, he or she will almost certainly have to be prepared to produce documentation for a subject about which their own knowledge is limited.

On the other hand, if the author is given the task of writing documentation without the input or aid of others who have technical understanding of the subject, then the author has to be able to adapt to short learning curves for a variety of technically related or unrelated topics, and be thorough in the understanding of the requirements of the documentation. Only then can the author present the information precisely and clearly.

READER KNOWLEDGE

It is unlikely that the documentation for a technical subject can be either interesting and/or useful to the reader if the author has failed to understand the reader's requirements. Knowing your reader is one of the most fundamental skills that needs to be acquired by a successful technical author.

Not making assumptions about the level of understanding of the reader is, perhaps, one of the hardest attributes to learn for the writer, since it is so easily overlooked.

Many examples of technical documentation show that writers are good at describing, say, a technical product, but few are able to offer the constructive information that tells the reader how to make the best use of it, or just to help the reader understand how the product works, or perhaps how it fits in with what they want from the product, etc. As an analogy, imagine you have to write an instruction manual for the use of a car aimed at someone who has no idea what a car is, or how it works. It is an easy task for the technical writer who understands exactly how the car works to describe all its features, components and what they do, but if the documentation goes no further than this, then the reader is unlikely to make much progress on making good use of the car.

Having described what a door is on the car, perhaps it would help to suggest which is the most useful door to use if you intend to drive the car along. It may sound silly in this analogy, but I can think of many similar situations occurring in software or hardware documentation where accurate but useless writing leads to time being wasted reading irrelevant sections of the text, before the apparently obvious to the writer becomes the same to the reader.

Understanding the reader also helps to structure the documentation logically and in the most helpful way. A detailed description of the workings of the carburettor and fuel system of a car is not necessary to help explain that the accelerator increases the amount of fuel fed to the car's engine, which is generally accompanied by increased velocity. But it is surprising just how easy it is for the author to fall into the trap of 'gassing on' (sorry about the pun) about something with which he or she is familiar, to the exclusion of the more relevant point being made to the reader. This is a lazy way to write and it takes good discipline for an author to cut through the waffle and find the thoughts of the intended reader.

ENGLISH SKILLS

It is fair to assume that the technical author should be a good communicator of English. A command of good English grammar is certainly an advantage. Some technical author training courses will suggest to you, however, that good English is not responsible for ensuring good documentation. This is primarily because few complaints are received about the grammatical peculiarities of a document from the reader when compared with the common complaints of 'I can't find the relevant section on XYZ', or 'This paragraph about how ABC works isn't right .... it doesn't do what it says in the manual', and so on.

However, I believe that while this may be true, good English cannot be ruled out as being one of the most important qualities of useful and respected documentation. After all, a well written and clearly presented piece of writing, which is easy for the reader to comprehend, must, by virtue, have been written with grammatical accuracy. Sloppy English can positively hinder the reader's concentration and any sentence or paragraph which has to be read more than once, because of poor sentence structure, is no more helpful than inaccurately presented facts.

Assessing the most important factors for good documentation

THE AUTHOR'S ROLE

is not an easy thing to do. There are a variety of opinions regarding this. There are those who regard factual accuracy for technical documentation as the highest priority. However, a technical specification of the electrical system of my car may be technically and factually correct, but doesn't provide the best documentation to help me find out how to change the fuse for the lights should it fail.

Another opinion considers that the most important ingredient of good technical documentation is that it should be well 'produced' — printed to a high quality specification of appearance, so that it becomes inviting to the reader. This idea is based on the assumption that if the documentation looks boring, the reader will avoid it unless absolutely necessary. If this is done at the expense of technical accuracy, the reader gains little knowledge at all.

Few will put quality of English grammar at the top of the list, but the fact is that, all these factors have their merits and their disadvantages. Good English certainly cannot be disregarded, and it is part of the role of the author to ensure that the information contained within a document is not only technically correct (whether or not this involves a contribution from a third party), but that it is also clearly written in a good style with few, if any, spelling errors. Any amount of production beyond this stage may or may not come under the authors' control, depending upon the areas of responsibilities assigned to them.

On the subject of spelling, I came across the course notes of a training seminar on technical authorship which implied that no-one ever complained to the publishers about bad spelling, and went on to qualify the need for technical accuracy first and foremost. Misspelling may not be of sufficient seriousness to provoke the reader into writing or telephoning to the source of the documentation to make a complaint, but apart from being mildly irritating, I, personally, find that bad spelling and punctuation indicate a lack of attention to detail, and this suggests that the technical content of the documentation could suffer from the same lack. If this maligns the quality of the rest of the manual, then it is the fault of the author who allowed such unnecessary grammatical mistakes to remain in the finished product.

ATTENTION TO DETAIL

The last comment brings me on to the next quality required of the technical author. This one, attention to detail, is often given as a requirement of the 'ideal' technical author. The good technical author is expected to be able to apply the same degree of concentration towards the details of the text as a book-keeper does to the journal and ledgers. Unfortunately, the job of being pedantic with the detail of documentation is often ignored by authors for two reasons:

a) insufficient time for the project to be done to such a level of detail; and

b) the disinterest on the part of the author to become involved in such mundane work.

It is certainly one of the least glamourous aspects of technical writing that, for example, indexes (or indices if you prefer) are painstakingly compiled, or that the sentence structure is examined more than once to ensure that the information has been presented in the best possible light. But the best authors are motivated by the desire to produce the best possible result. The determination to succeed in this task, therefore, must involve concentration and attention to detail and these attributes should figure high on the list of author qualities.

COMMUNICATION SKILLS

The technical author is, by the very nature of the job, a communicator; a communicator of technical information to both technical and non-technical audiences. However, it is in another direction that, as far as communication goes, the author must possess another 'ideal' trait.

In many cases, technical writers are introverted individuals who are at their best working in isolation. They do not necessarily receive the maximum job satisfaction when in a team. By the nature of this personality trait, which does seem widespread, they are not the best communicators with other individuals. This creates two major problems. One is that they have to communicate well with others to access the source information for the documentation, and to

THE AUTHOR'S ROLE

successfully convey information to others, their peers and managers, about the projects on which they are working.

The second problem is that this manner of personality can be regarded as being partly responsible for the lack of acceptance that the author receives, in being given equal status to, say, marketing or sales personnel within an organisation. If the author can communicate well, it will assist in the success of achieving the goals of the job.

Often the author falls between two distinct camps within an organisation — the technical and marketing departments. As such, authors often find themselves acting as the bridge between the two. The technical departments may not communicate well directly with the company's marketing forces, since they do not talk on the same technical level. The marketing divisions often don't understand (and don't want or need to understand) the technical complexities of a product. However, they do need to know the important and salient features of a product if they are to target its market position well, and so that the company can provide adequate technical support for sales, after-sales support, and problem solving. The author can often provide the link between the two departments.

An author who can easily communicate with technical people but also recognises and understands the needs of the novice or non-technical reader, often has the ability to convey the technical qualities of a product to the less technical marketing individuals. This is presumably the reason why so many technical authors become involved in writing sales promotion and marketing literature, particularly where product descriptions are involved. They may, for example, supply the facts and technical details to the copywriters of product or technical service brochures, who then add the usual flair to the text which gives it that 'zappy' and hard-selling approach that the sales promotion people love to see.

So communication is important; communication with individuals of the same organisation as well as communication with the reader.

LESS APPRECIATED QUALITIES

All of the above qualities are offered as the most likely requirements of the ideal technical author and help to describe his or her role in the organisation. However, there are less obvious qualities that form part of the author's role and assist in their success.

The author who understands some of the qualities and skills of good management will be at an advantage. This is because, in the smaller organisations, authors are not managed well. In particular, the company that employs the 'all-singing, all-dancing' author-manager, whose tasks include the writing *and* producing of the documentation entirely by him or herself, is usually poorly managed by his or her superiors.

This is because, in general, managers do not sufficiently understand the role of such an 'animal'. In such cases, author-managers must manage themselves. This involves paying attention to such things as:

a) time management — efficient control over the use of one's time;

b) management reporting — to superiors, including scheduling, productivity control, and efficiency analysis;

c) economic control — careful attention to costings, pricings, etc, and the sourcing of cost effective production methods.

Looking for a way to improve one's own environment should be regarded as fairly high priority. Don't wait for others to do this for you. Assessing oneself from a time analysis point of view can help you to increase your own output. If you find, for example, that you spend x hours on a job making enquiries of technical or development staff about each new product, when they themselves could provide some specification documentation in advance, you may be able to persuade them to help cut down both departments' time in the long term.

A FEW QUESTIONS TO ASK YOURSELF

To improve your performance, and produce better results in anything, you have to analyse the way you do what you do, and ask questions about the way you go about your job. This is not unique to technical authors, of course, but below are some questions you should pose yourself, if you haven't already considered them:

THE AUTHOR'S ROLE

- How much do you know about how cost effectively you produce documentation when compared to others?
- How well do you know the competition?
- Have you taken the time to examine documentation from other sources on a similar subject, and, if so, can you recognise the good and bad qualities of it? Most important of all, what can you learn from it?
- Are your resources adequate, and can you properly justify the need for additional resources to your superiors?
- Do you work to a budget, and, if so, do you personally control it?
- How successful are you at scheduling documentation, or do you rely upon the scheduling of a technical development department and then just add some guesswork to it?

These questions should highlight to the author-manager where some of their less obvious responsibilities lie. The publications manager will take care of many of these aspects, if one is employed in such a capacity. If you work for one, then he or she should already be addressing these questions. But the author must appreciate these requirements if he or she is expected to do more than just write the words. Remember that the author-manager has a responsibility not only to the reader of the documentation, but also to the company he or she works for, such that the company is providing, where required, the kind of documentation which is both well received by the reading audience, and efficiently and effectively produced. In addition, the documentation that is distributed outside the company should satisfy the need of adequately projecting the company's profile in the market place. More details about the role of an author-manager can be found in Chapter 9, The Publications Manager, since many of the areas of responsibility overlap.

IN CONCLUSION

The role of the author, then, is a varied one. In particular, one should be aware that the way in which the author is regarded in his or her own organisation can affect their success or failure in producing high quality technical documents. If the author is viewed as simply a technical-minded individual who writes about how a product works, or how to fix faults, etc, then this is no more a comprehensive understanding of an author's role as suggesting that the role of a salesman is simply that of selling products or services. Such is a very limited description of the salesman, who invariably has many responsibilities that include maintaining the loyalty of customers; providing appropriate reassurance in the event of problems; acting as a consultant and advisor on behalf of the company; being responsible for the forecasts and budgets for future sales; the increasing of sales potential to get more job satisfaction; and the improvement of the company's profits.

All these qualities are more easily recognisable in the commercial sense, but many of them, perhaps surprisingly, overlap into the kinds of responsibilities of the successful technical author in commerce, whose documentation carries with it the company's image to the reader. But few would include such items in a list of an author's responsibilities, especially those who are *not* technical authors. Ask your manager to describe your job responsibilities to you and see how many areas are covered. You will probably find the response falling somewhat short of what you think. On the other hand, if you work in a large organisation that includes a department of technical writers, controlled by a manager whose background is associated with technical documentation, then you are more likely to receive a sympathetic understanding of your role.

If the individual author wants a little help in communicating his or her status to others, then he or she must realise their own responsibilities to their employers. To those authors who have told me that, because they are on their own, they do not have any 'weight' when it comes to getting financial or other resources brought to bear on helping with the workload, or that they are not adequately recognised or rewarded for their status and abilities, then, brutal as the advice may sound, I would say you can help yourself if you stop thinking about your own problems. Instead, concentrate on providing information to your superiors about your workloads, resource scheduling, and cost analyses.

If you think you don't have time, you have an immediate resource problem anyway, and this should be discussed without delay. Check

THE AUTHOR'S ROLE

these following points, and see if you honestly satisfy yourself that you are doing all you can:

- If you have to attend meetings, then be armed with information.
- If you don't already schedule your work, do so.
- If you do not have a cost analysis of how your last 10 or 20 documentation jobs worked out, start analysing now. Have information to hand, and supply this to your superiors.
- Allow your superiors to see how your job affects the overall company plan.
- Offer ways in which you can improve upon what figures you have provided, and try to incorporate the requirements of others when it comes to meeting deadlines.
- Try seeing the point of view of others.

On the whole, your superiors will respect you for your efforts, and view you as being more of an integral part of the success of the company's plan than may have at first been appreciated. You should find that the problems of recognition and resource allocation will become resolved gradually and naturally. If you are still not convinced, I offer such advice on the basis of personal experience having done just that, and increased the resource budget of the documentation department by over 150% in the space of 18 months, while reducing the cost per page of producing technical documentation by 275%, yet still continuing to improve on the quality of materials used and the presentation and packaging.

If you are still concerned about the time you need to provide such information, remember that this does not need to occupy too much time, since it does not need to be done in great detail, initially, to help the situation move in the right direction. Fuller and more comprehensive control can only be undertaken if you have adequate time and resources, and in any case, such matters would normally be dealt with by a publications manager. If your role as author overlaps, then these areas begin to fall into your area of responsibility and you mustn't ignore them.

Fortunately, the role of the author is becoming more understood by the industry as a whole. The status of the author is generally being improved as time goes on and the recognition of this is being reflected in the salaries offered throughout the computer industry, in the jobs advertised for author positions.

Freelance authors are regarded as quite high status jobs, and attract the experienced author as much for the variety of work, which can aid job satisfaction, as the freedom the author can exercise over his or her skills. Contract authors hold a similar position and since this is the area into which some of the best experienced authors are beginning to move, then the reliance upon contract authorship remains attractive to many of the companies that need technical documentation services. So, if you feel you're hard done by as an author now, the future could hold better prospects for you yet!

2 Writing Sources

Input to technical documentation within the industry comes from a wide variety of sources. Not all documentation is, by any means, the exclusive work of a single technical author. In some cases, the technical author may be only one of many resources involved in the writing, and input may be so significant from other parties that the job of the author becomes more like that of an editor.

When producing user documentation for a technical product, for example, the technical or development departments may provide a significant amount of specification material which constitutes the groundwork of the user documentation itself. In the software industry, a well written specification for a product can certainly cut down the amount of time that the author has to invest in finding out the software's features, its input and output limitations and the way it processes information.

In the production cycle of a technical document, it may be that a working team is created to cover various aspects of documentation, which includes individuals from more than one area of specialisation. Such individuals may include training staff, whose job dictates that they must understand the subject of the documentation in any case, and may well provide useful input to the project which can be of considerable help to the author. After all, much technical documentation is intended to 'train' the reader in the understanding of the subject matter, and an experienced training officer can often provide the kind of insight required to improve the way in which documentation puts across its meaning to the reader.

In a national survey of documentation for the computer industry, organised by Digitext, an analysis was based on a question which asked for the percentage of involvement for each resource used to write documentation of varying types of project. The responses to the question were treated as instances of a writer's involvement expressed as a decimal, relating to the percentage, eg 1 for 100%, 0.5 for 50% and so on. Table 2.1 shows the results of the analysis.

	Technical reference manuals	Installation manuals	Training manuals	User manuals	Intro. manuals	Other documents	Total
Product developer	5.40	4.85	2.00	7.20	1.60	1.05	22.10
Project manager	1.40	.10	.10	2.10	.70	0	4.40
Trainer	.20	.20	11.45	.60	.40	0	12.85
In-house writer	10.30	7.25	9.80	14.05	8.70	5.35	55.45
Contract writer	.20	1.10	0	1.45	1.60	0	4.35
Documentation house	.90	.50	.15	4.30	2.00	2.60	10.45
Other	2.60	2.50	.50	1.40	1.90	3.00	11.90
Total	21.00	16.50	24.00	31.10	16.90	12.00	121.50

Table 2.1

Digitext clarified that the 'Other' writer category included engineering, support staff, marketing department and educational services, while the 'Other Documents' category incorporated brochures, reports, fact cards and data sheets, service manuals, system specifications, quick reference guides and so on.

The analysis showed clearly how much involvement belonged to the technical authors, while the product developers came in second place. In conjunction with this, the survey also analysed, in a similar way, those responsible for reviewing the draft documentation with the roles reversed — development staff playing a more significant role than the technical writing department. But the details show how many people can be involved when a document is being produced and how the main source of writing can vary from one project to another, as well as within organisations themselves.

USING OUTSIDE SOURCES

Using third parties for writing sources is still regarded as a valid means of producing technical documentation, particularly where user

WRITING SOURCES

manuals are concerned, and though there were no specific reasons brought to light in the survey why user documentation should be singled out as being acceptable for writing by a third party, one of the factors is likely to be quality.

Most companies consider that the quality of *user* documentation requires the greatest attention. In a complementary question in Digitext's survey which asked what reasons companies had for using contract writers or a documentation house, the reasons given could have been applied to any of the types of literature mentioned previously. They included (in order of priority):

a) High workload
b) Specialist skills required
c) Urgency
d) New documentation/product release

and only one respondent gave the reason that there were no in-house capabilities.

For those companies without in-house writing capabilities, the services of either a contract author or documentation house can be attractive, particularly if the recruiting or appointment of a writer would not allow sufficient time to complete a documentation task already in hand (having considered learning periods, for instance).

Even for those companies which already employ authors, internal resources may be stretched to a point where, for a one-off job, further expansion of personnel in this direction would be uneconomic, or perhaps a specialised skill may be needed for a particular job, which is not available within the organisation. In this type of situation working with a documentation house as opposed to a contract writer can have benefits. Often the documentation house can provide an author most suited to the task, with appropriate background, experience or knowledge for the project in hand. The author is then project managed by the documentation house who can monitor progress, and check and assess quality on your behalf.

Whether you use a contract writer or documentation house, there are certain key points that require your attention, and a few dos and don'ts about third party authorship. Here are seventeen points which you must consider when thinking of using third parties:

a) When choosing a contract writer or documentation house, find out to what extent their experience in this field is relevant.

b) Ask to see examples of previous projects on which they have worked.

c) Get the author or project manager to visit your site to assess the task you require of them, and to see your set-up.

d) Find out exactly how they intend to manage the task, for example, if the author is going to be working remote from your site, is this going to cause communication difficulties?

e) Make clear to them how you intend to help. Will you be supplying them any information, or is it up to them to find out?

f) Clarify whether you have to supply the third party with any form of equipment for them to carry out the task. If so, where will the equipment be taken, for how long, and who is going to insure it?

g) Make sure proofing cycles are clearly defined, particularly where you or other individuals are expected to be involved in the proofing. To what extent are you responsible for the accuracy of the text? In most cases, the contract writer or documentation house will endeavour to submit proofs or drafts which you will have to approve and for you to suggest changes. In these circumstances, you may become responsible for the finished product, even if it ultimately fails to fulfil its purpose — check the small print of their contracts.

h) Look for any seals of approval. If the third party quotes other clients, try to establish contact with them and ask them if they were satisfied. If there is any association with organisations such as the ISTC (Institute of Scientific and Technical Communicators) contact them for comment.

i) Make sure that you are clear about charge rates. If the charges are to be on an hourly basis, estimates for the job must include any additional time required for proofing, amendments and re-proofing.

WRITING SOURCES

j) If the author has to work close to or on your premises for the purposes of resources required and/or for ease of communication, you will probably be responsible for costs and expenses incurred in travel, accommodation, etc. Don't forget about them and get a quote for such expenses from the individual or documentation house.

k) If timescales are specific from your point of view, or deadlines have to be met, consider discussing what contingencies there are if the author goes sick, changes job or simply lags behind — protect your interests where possible.

l) Ask for the details of time estimates, costs, expenses, etc to be provided to you in writing, after the job has been assessed. As the job progresses, ensure you are provided with timesheets detailing the work on the job, so that you can monitor progress against these estimates.

m) Find out in what format you are to receive the texts. Are you or the third party responsible for word processing, coding for setting, typing or whatever? Will the text be supplied on disk, and if so, on a disk format that you can use? Perhaps the documentation house will handle print production for you as well.

n) If illustrations are required to assist the text of the technical documentation, find out who will be responsible for them, and, if the third party is providing illustrations, will they be suitable as artwork or will they require additional resource to make them camera ready, for example.

o) Establish how the charging is to be done for the job. In some cases, where a job may span several weeks, the contractor or documentation house may require payment against timesheets submitted at intervals throughout the life of the job, rather than invoicing you once it is complete. Find out which method will be used, but remember that there are two considerations here: a step by step method of invoicing spreads the cost for you, more like paying the salary of an author you might employ; however, you will have less leverage on the writer or documentation house if problems occur towards the end of the process and the job is still not complete.

Careful monitoring of an open-ended charging structure will be required to ensure that a job quoted for at £8,000 doesn't suddenly creep nearer £10,000 as the 'unforeseen additional workload' increases as the job progresses.

p) Consider getting more than one third party to quote on the same job.

q) When a quote has been received, consider the alternative of how far the cost will go towards the salary of an in-house author and whether you could employ the time of an in-house author beyond the immediate task requirement.

If all the above seems to put you off using an external source for documentation, don't forget these advantages:

— for a short term project, investment in additional resources may be costly if you cannot employ that individual's time after the project is complete;

— a contracted author may have specialised in the area in which you are seeking skills, and therefore have a broader knowledge of the documentation requirements in his or her field than an in-house trained author;

— you may consider that using a third party makes your budget control on the project easier than can be established internally;

— some documentation houses can provide contract authors who can integrate with an existing team, bringing a fresh approach to the style and format of your existing documentation;

— it is in the interests of single freelance authors to complete jobs on time for two main reasons: they rely upon your satisfaction with the work to gain recommendations for further work, and they need to ensure that they have a job to follow on — if one is already lined up, as far as they are concerned, they cannot afford to overrun on a current job.

This last point is significant. While third party authorship is almost invariably more expensive on a pro-rata basis, compared to the costs of a full-time employed author, external services can often work more efficiently than internal resources, since they have a more clearly defined commercial interest to perform efficiently. This emphasises

WRITING SOURCES

a point I made in Chapter 1 about an author's role: that of self discipline and time management.

COMMUNICATING WITH INTERNAL SOURCES

Assuming you have established who is going to be responsible for taking on the author role in your technical documentation project, the next consideration is the way in which communication channels are set up between the author and others who contribute to, or require contributions from, the documentation cycle.

If technical staff are involved in providing data, proofing drafts and so on, you should define a mechanism by which their contribution can be controlled in the schedule of the project. Documentation jobs are often held up at draft or proofing stages where other individuals are required to check or amend a typescript. After all, people generally consider that their own work is more important. Typescripts are therefore either shelved until those individuals get around to checking them, or less than the minimum degree of attention is given to the job, and the draft document speeds past the eyes of the individual concerned to get it out of the way. If this situation sounds a little familiar, you should try to organise a second line of checking where possible.

Setting up project-based meetings on a regular basis can be a good way of monitoring the progress of the job. In these circumstances you should arrange to involve all parties who contribute to, or are affected by, the production of the documentation. Communicate with everyone — don't forget those people whose job it is to dispatch the documentation, for example, they need to know when it's going to be ready! Consider providing data sheets on the current levels of documentation, and if it is complex and made up of several components, a kind of 'parts list' does not go amiss in helping other departments to understand what comprises the current levels of documentation. I have heard of, and experienced, many instances where old documentation is being dispatched to clients, while more up to date versions sit on the shelves in stores, or unprocessed in the print department.

If you are using external resources for print production, design, packaging, etc, representatives of these areas of responsibility need to be kept informed throughout the process. If more than one source

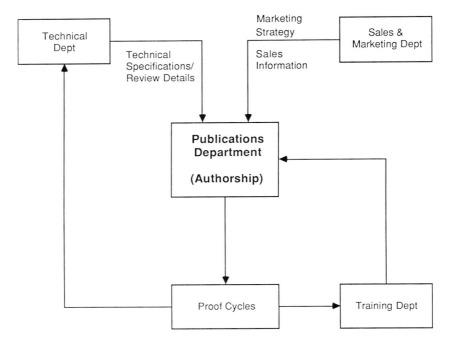

Figure 2.1 Writing Source and Input Controls for a Documentation Project

for writing is involved in any one project, try to get into the habit of sketching out who will do what before you plan your schedules. This simple task will help you to establish a clear picture of how the whole project fits together. For example, Figure 2.1 shows a diagram (sometimes referred to as a 'plannogram'!) of the writing source and input controls for a documentation project. By formalising the structure of each project this way, you keep a 'picture' of the task in your mind's eye; this helps you to control the documentation process, and ensure that all involved are kept informed of the progress.

ASSESSING THE COMPETITION

It is good practise to be aware of how your competitors in the industry produce their documentation. Whenever you consider other documentation in this way, you should always consider yourself to be

WRITING SOURCES

in a 'learning mode'. It is easy to simply criticise other documentation for the sake of it, in order to satisfy your own self esteem; be critical where it is due, by all means, but recognise why your criticisms are formulated, not just what they are. Be prepared to learn from what *you think* is a good feature in someone else's documentation, and copy it. Chances are, if it is a good idea, they got it from someone else anyway.

I stress that you learn from what you think is a good feature because such things are often speculative, and if you asked ten people what they thought about one piece of documentation, you'd probably get ten different opinions. It's up to you to learn as much as you can from other examples, and in this respect, other documentation which has been produced to fulfil a similar role to your own is in itself a 'writing source'. It provides input from a critical point of view, and offers you the opportunity of getting a clearer sense of whether you are satisfied with your own efforts or not. Simply reviewing similar documentation provides a measuring rod for your own.

IN CONCLUSION

To summarise very briefly, you should be aware that there are many possible writing sources. Make the best use of the resources your company has, and communicate with other departments. Learn to identify where a useful source of additional information may lie, like the training department, for example. Involve those who know the subject matter in the proofing cycle. Remember, as well as receiving information to make your task easier, consider giving information to others to make their task easier too. Look out for, and learn from, documentation from other sources which fulfil a similar requirement.

Here is a checklist:

a) decide which resources are best suited to control the authorship, internal or external;

b) specify all parties who are contributory to the documentation process, identify your sources of information, proofing cycles, etc;

c) monitor the progress of the project, informing and consulting all listed in b) above;

d) measure your documentation against others which have a similar purpose.

Some further information of third party documentation specialists can be found in Appendix A.

3 Format and Style

All technical documentation tries to 'teach'. The differences are between those that teach well and those that teach badly, and between those that teach valuable things and those that teach irrelevant or trivial things. An author who does not care about how he or she writes is just as likely to be a bad author as one who does not care about the subject being 'taught'.

Good communication through documentation, like good design, begins in the mind. Certain attitudes and thought processes will precede the words, and a good communicator will always consider the recipient of the message. So must an author consider the reader when forming the text through which the message is conveyed. The form and style adopted in the writing must be suited to the purpose of the documentation, and not simply adhere to rules or regulations laid down by some preconceived notion about technical documentation. Except for cases where documentation must conform to some British standard or specification, the author's job includes the responsibility of considering the most suitable form and style.

This is not to suggest that there are no basics or principles on which the art of technical writing can be taught, since there are recognised, tried and tested approaches which should be learned, and on which individual style can be built. Since most writing of any nature begins with the language, it is in this area that you should examine your skills first.

GENERAL USE OF ENGLISH

Writing began as a specialist skill. Today, *good* writing, particularly good technical writing, is also a specialist skill. Since becoming a widespread form of communication among all walks of life, in an overwhelming variety of forms and styles from academic literature to gutter press, the English language has come under tremendous stress and undergone, and survived, considerable abuse. In industry and commerce, where new technology and progress are stretching the vocabulary of the English language, we see newly invented terms and expressions, jargon and 'empty' words being invented at an astonishing rate.

Sadly, technical writing, more than any other, could be most criticised for encouraging that trend. That is not to lay the blame solely on technical authors, but the degree to which writers of technical material have contributed to misusing the grammar of language is significant. This has been due largely to two major factors.

Firstly, those whose task it is to document a technical subject tend to have, more often than not, a more thorough understanding of the technical subject itself than of clear and precise communication. They will have been used to writing papers for journals, technical reports, summaries and the like, most of which are aimed at a readership of similar background.

In the computer industry, it is still commonplace to find the technical authorship of user documentation has been carried out as a secondary function of the programmer or technician. Some of the worst examples of documentation of software products are so obviously penned by a programmer that you could almost assign sequence numbers to each line of the text.

The second factor responsible for giving technical writing a bad name is the continual use and invention of jargon. Technical writers (and communicators in the technical world as a whole) have an uncanny ability to invent it faster than any other user, or rather misuser, of the English language. According to Fowler (1968), jargon is:

> 'that "talk" that is considered both ugly-sounding and hard to understand: applied especially to (1) the sectional vocabulary

FORMAT AND STYLE

of a science, art, class, sect, trade, or profession, full of technical terms; (2) hybrid speech of different languages; (3) loosely the use of long words, circumlocution, and other clumsiness.'

Unfortunately, though perhaps not surprisingly, it is in sense (3) that technical writing finds itself responsible for the increase in use of jargon words and phrases.

The worst offenders must surely be those who populate the computer industry. Here one finds such ugly words and phrases as 'abort the function' (meaning to cease processing); 'escape returns control to the menu' (meaning press the escape key to re-display the menu); 'kill a file' (instead of remove or delete); 'dynamically expanding file structures' (files that get bigger the more data you add).

People who work in technical environments will encounter many technical words, of which only a few are intelligible to the layman. It is essential, therefore, to determine the appropriate level of vocabulary to be used when communicating through your documentation. So often, technical jargon is used where more common words are adequate.

There are countless examples, and those authors who work in a high technology environment may find it difficult to remain constantly aware of the specialised nature of the terms that they use in the text. Many users of computers and software will have come across the term 'format', as relating to the formatting of a diskette, but how well can you explain the term to a user who has never used a computer and who is reading some documentation for the first time? It is an example of a term where many of us know what it means, but may find it difficult to explain in words. We all *know* what light is, but try *telling* someone what it is.

Looking in a glossary for a user manual for computer software, I checked on the word *format*. It explained that to format a disk is a process which initialises a disk for use on the computer. I was none the wiser, or at least wouldn't have been had I not formatted thousands of disks in the past and knew what happened if I didn't do so when it came to putting it in the computer's disk drive.

The lesson to be learned here is not to become complacent, and don't assume that you should treat technical documentation aimed at a technical audience with any less respect to clear writing as that aimed at a novice readership.

Be ruthless in your use of accurate and precise English. But write simply. Whilst you are responsible for communicating good English in every sense, one can easily fall into the trap of being wordy (verbose circumlocution to be wordy about it!). Use a thesaurus, but only to find simpler ways of saying something. You don't need to impress the literary dons at Oxford, you are trying to communicate instructions, as quickly and as plainly as possible, or convey ideas and information to readers who are using your documentation, probably because they have to, not because they want to. You get no extra points for making your text flow with literary excellence. Leave that to non-technical authors, or write yourself some great literary work if you're inclined to do so. In the meantime, concentrate on the following word and act on it:

SIMPLICITY

Let me illustrate this with an example. The following passage of text was taped to the side of a PC screen by a technical author who once worked for me, and it served as a constant reminder about the need to keep things simple:

'Such preparations shall be made as will completely obscure all Federal buildings occupied by the Federal government during an air raid for any period of time from visibility by reason of internal or external illumination. Such obscuration may be obtained either by blackout construction or by terminating illumination. This will, of course, require that in building areas in which production must continue during a blackout, internal illumination be provided, that construction may continue. Other areas, whether or not occupied by personnel, may be obscured by terminating the illumination'

Re-write by Franklin D. Roosevelt:

'In buildings where work will have to keep going, put something across the windows. In buildings where work can be stopped for a while, turn out the lights'

POINTS TO CONSIDER

When considering the format and style that you should adopt for your

FORMAT AND STYLE 37

technical documents, and indeed these may change from one type of document to another as well as from one organisation to another, there are common points to which you should direct your attention, and these are detailed below.

Completeness

Technical documentation must be complete, and specific about what it is to convey. If you are documenting a fault finding procedure, consider the questions that may be asked by the reader when trying to establish what action to take, rather than simply listing answers to problems.

Take the text outside the area of a simply descriptive technical work like a specification, to consider what the reader is trying to achieve from the documentation. For example, if a technical document is intended to describe the installation procedure of a piece of equipment, apart from covering all the tasks necessary for installation, think about what action may be taken if the results are unexpected.

Here is an illustration of that process. The following paragraph is a hypothetical installation routine documented for an imaginary item of hardware. For the purpose of this example, we will assume that we do not have the benefit of an illustration (though you might normally include one).

'Remove the system components from the packaging and place on a flat, even surface. Connect one of the power cables supplied to the power socket at the rear of the processing device, and the other to the power output of the VDU. The cable connected to the VDU should then be connected to the port marked "video" on the processor. The keyboard connects to the socket marked "KB" on the rear panel, next to the processor. Switch the system on. The device will start a warm-up procedure during which self diagnostics will check that the system is correctly functioning. Any errors will be reported on the VDU during this process. Assuming that all is working well, the screen will display a message "insert boot diskette into drive. Press any key when ready". Follow these instructions, and wait for the system menu to be displayed....'

The above paragraph, although a very simple illustration, may be technically accurate, but does not tackle the documentation of the installation in the best way from the reader's point of view, except in the circumstances where, by chance, everything that the reader needs came with the packaging, and that everything went smoothly according to the instructions. Now see whether you think this is any improvement:

'You will need a flat, even surface to assemble the components of your system, and access to two power sockets nearby. No tools are required for the assembly.

Remove the system components from the packaging. You should find the following items:

Processor device
VDU (with cable attached)
Two power cables
Diskette labelled "Boot Disk"
Keyboard (with cable attached)

If anything is missing, report the matter to your supplier immediately.

Connect the components' cables as follows (all connections are made on the rear panel of the processor):

One power cable into the power socket of the processor, the other power cable to the socket on the VDU. Plug the free end of the cable attached to the VDU into the video output port on the processor (marked "Video"). Plug the keyboard's cable into the socket marked "KB" on the processor.

Make sure the power cables are then connected to a mains supply before switching on the device.

Once switched on, the device should begin its warm-up procedure. You will hear the cooling fan at first, then a beep, as the start-up process begins. If nothing happens, there may be a fault with the equipment, the cables, fuses may have blown either on the power cable plugs or in the rear panel of the device, or you may have simply omitted to switch on the mains supply....'

and so on.

The essential difference between the first version of the installation instructions and the second, is that the latter version thinks about

FORMAT AND STYLE

the reader and what he or she is trying to achieve, and what action may be required if things don't go according to plan. The very first sentence of the second version considers what environment the equipment is to be used in, and may easily escape the consideration of the author of the instructions who already knows what environment the equipment is being used in, and is busy documenting the exact stages that he or she would go through to start the system up, and nothing else.

A little thought about mentioning what noises the equipment will make when it is switched on might perhaps comfort a user who is totally unfamiliar with such equipment and might be concerned at the sound of a beep or two, which could suggest a problem rather than a natural course of events. This is all over-simplified, but gets across the point that completeness in any form and style of technical writing is important, and makes no assumptions about the reader's level of knowledge in this example.

There is no case to argue that you may be condescending to a reader who might consider some of the instructions or text familiar. For one thing, a reader who comes across areas of text covering points about which he or she already knows will simply skip them. From a psychological point of view, there can be some satisfaction on the part of the reader that at least something in the documentation relates to something with which they are already familiar, and this creates a sense of making the reader comfortable with the text. If the first paragraph, and indeed every one following it, in a technical document contains only material which is new to the reader, one can feel a little daunted at the task of reading the rest, particularly if it is a 400 page manual.

Relevance

Having said that documentation should be as complete as possible, make sure that the content is relevant. There is little point in providing the reader of the installation instructions illustrated above with a technical breakdown of the fault diagnostics of the machine as it starts up when, if any fault arises, he or she is going to have to call an engineer anyway.

Including such irrelevant, if accurate, material serves to disrupt the flow of reading. If such information is required in the documentation for reading by someone other than the person who may be referring to the current section, provide such information in an appendix and include a cross reference to it.

When designing your documentation, consider how best to present information which is relevant to any individual. For example, if user documentation for a software product describes how the operator is to input information, and details of installing the software, intended to be read by a supplier or technical installation engineer, are also required, consider supplying this information as two separate documents.

This way, the operator feels there is less documentation to wade through than may at first appear — always a good psychological move for users if you want them to read it — and the person responsible for installation may require details that need more frequent updating than the remainder of the documentation. If such detail is separate, your task of revising the documentation can be simplified administratively.

Ambiguities

Avoid these at all costs. Get someone else to read your draft documentation and see if they understand what you are trying to convey to them.

Consistency

If you do not already have one defined for your company's documentation, a specification for house style may be a good idea, to maintain consistency in the format and style of your technical documentation throughout. This is particularly important if more than one resource is used for writing.

If you need the services of a contract writer or documentation house at some point, it will be useful to be able to provide them with guidelines about the format and style of your existing documentation, so that it remains consistent. To do this, you may have to formalise your styles and may not have even considered this important. To help you with this task the following section discusses the specification of a house style.

FORMAT AND STYLE

HOUSE STYLES

Unless you are working on technical documentation which has to conform to British standards, or some other standard or specification laid down by appropriate authorities (particularly relevant for documentation of government and military associated projects), you may not have considered the need for defining your house style. By doing so, it helps you to consolidate the format and style used throughout your documentation such that a set of rules can be clearly defined for any author or contributor to the text to follow.

The following are some of the aspects of your documentation specification that you may wish to consider.

Page Style

Determine the layout styles to be used for technical documentation where it is practical to adhere to a format (see also Chapter 4 on design and layout). If you typeset your documentation, this detail should include measurements for line width, typestyles and point sizes for text, headings, sub-paragraphs, highlighting, etc.

Headings

Specify how sections or chapters are to be headed. Will new sections/chapters begin a new page? What kind of section numbering should be used, including sub-sections or paragraph numbering, and to how many levels?

Spelling/Hyphenation

Do you have a preference regarding spellings? For example, specialise or specialize, organisation or organization, indexes or indices? (the use of the z is *not* an American format, incidentally). For hyphenation, would you use sub-division or subdivision? Whatever you choose, keep it consistent throughout.

Punctuation

Specify any peculiarities that you would require in this respect, for example, it is generally accepted to use only single quotes, reserving double quotes for quoted words within a quotation.

Capitals

Often used in a very haphazard way in technical documentation. For example:

'This sets the control of the Request To Send signal. If set to OFF, the RS signal goes ON when the device is outputting data.'

Here the use of capitals on 'Request To Send' and for keywords like 'OFF' and 'ON' may have significance, but again, a set of guidelines needs to be provided and then adhered to.

Highlighting

Specify when to use text highlights such as underlining, emboldening or italics, etc.

Abbreviations

Decide whether abbreviations are to be used with or without full stops, and keep abbreviations consistent, eg MB for Megabyte, Mb for Megabit.

Dates

Specify the format to be used for dates, and how they may be abbreviated. For example, 19th July 1989, July 19th 1989 or 19-07-89.

Units

If you are using SI units, make sure that you use them according to standards, for example mm, m, km (but not cm), g (not gm), etc. Abbreviated units do not normally change in the plural, so use 3 km (not 3 kms) and so on.

Notes and Reference Systems

Specify how footnotes and references to other publications, technical reports, journals, etc are to be handled.

Tables, Lists, Figures, Maps, Plates, etc

Decide how these are to be presented and referenced within the text.

FORMAT AND STYLE

For example, you may wish to refer to the first illustration in section 3 as fig. 3.1, the second as fig. 3.2 and so on. Similarly, tables may be referenced as Table 3.1, 3.2, etc.

For lists, decide how they are to be numbered, if at all. Do you prefer a), b), c) or 1, 2, 3? Consider also the indentation of lists and follow a consistent rule over the use of capital or lower case letters at the beginning of each item, and whether a full stop or semicolon be used at the end.

Copyright and Trademarks

If you quote from other works, you may need to acknowledge the copyright, and all trademarks may be acknowledged in the text or among the leading pages of the document.

Indexing

This is always a difficult area, but try to establish what criteria should be used to decide whether an item is to be included in the index or not, and how it is to be presented. For example, 'RS232 serial interface' may be included in the index under R for 'RS232 serial interface', or under S for 'Serial interface — RS232' or P for 'Printers — RS232 serial interface', or all of these.

Once you have specified your house style, get it typed or word processed and distribute copies to all those concerned. It also makes the proofing of documentation more formalised.

INCLUDING ILLUSTRATED CONCEPTS

For all their importance, words must not be confused with the things that they represent. Words are only symbols. They are not always the best medium for communication. A picture can often impart information instantaneously in a situation which would require skilful writing and a lengthy reading time.

Technical documentation often includes illustrations for the purpose of communicating design information. But using illustrations to portray a concept or to give a summary of the points covered in a chapter can not only help convey the meaning, but maintain the interest of the reader.

STRUCTURE OF THE TEXT

Apart from acquiring the information that you need for the body of the text itself, you should consider its order and presentation. Information that is written to convey understanding of a subject must be done so in a logical fashion.

Structuring the documentation before beginning the writing process can be a considerable advantage to the author, as it enables him or her to get a sense of how the various components relate to the documentation as a whole. I have found that, particularly for documentation that is not or cannot be written in any sequential way, ie from page 1 to the end, the first thing I try to establish is a list of contents. Once the contents are clear, any section or chapter can be written independently, in any order, since it is easy to see where it fits into the logic of the documentation's structure.

IN CONCLUSION

To round up this chapter on form and style, Table 3.1 is a check list for the author regarding the proofing procedures for ensuring form and style is given adequate attention:

Stage	Nature of review	What you should check for
Draft	Logic/Structure	Subjects in correct order Duplications Irrelevancies Ambiguities
Amendments	Clarity/Accuracy	English Technical content Jargon
Editing/proofing	House Style	Consistency Adherence to standards, specifications, house styles
Final proofing	Everything	Ease of reading Illustrations Cross referencing Indexes Flow of reading/ideas

Table 3.1

FORMAT AND STYLE

If you pay as much attention to form and style as you do the technical accuracy of the content of documentation, you will be well on the way to improving the acceptance of your technical offerings in the eyes of the readers. If you want to catch their eye so that they feel they want to read the document in the first place, you must consider design and layout, and these subjects are covered in the next chapter.

4 Design and Layout

INTRODUCTION

Successful presentation of technical documentation relies on good knowledge of the basic principles of layout and design. The purpose must be to give a visual form to the work by arranging the elements of typography, illustration and space in such a way as to communicate the content of the text in the most effective way.

By concentrating some effort in this direction, many technical documents would be received with a far warmer reception than they generally do. For those non-technical readers who have to read technical material in order to understand the working of some system or product, for example, the mere thought of reading material of a technical nature can be ominous. Therefore, it is the duty of the publisher of such works to provide some encouragement. The layout and design should aim to do at least the following for the reader:

a) invite the 'opening' of the documentation in the first place;

b) give the impression that the contents are easy to follow;

c) encourage reading of the text; and,

d) help the reader find what is wanted.

If the design and layout of the document fails to do the above, then no matter how accurate or well written the text, the document will not serve its fundamental purpose of communication.

LAYOUTS

It should be understood from the outset, that design and layout cannot be simply a matter of guesswork if it is to work well. With the advent of desktop publishing systems, the tools for layout are now in the hands of many who have not been trained in the principles of typography and design. One can easily be drawn into using features in these DTP software packages without considering the implications.

Take, for example, the ability to set text for documents and manuals in multiple, justified columns. This may be a novel way of approaching documentation if you have never before had the ability, in-house, to generate layouts in more than one column, but unless you understand the purpose for using column layouts, you may not be doing your presentations justice. More about desktop publishing and the perils of its use in unskilled hands is covered in Chapter 7, but to continue with the multiple column example, consider the use to which the final document is to be put.

Technical training notes, for example, may benefit from a single column of text with a wide margin and hanging headings. The headings make the sub-sections easy to find, and the wide margin serves as useful space for penning notes and attaching comments to the text (see Figure 4.1).

The best layouts are often produced from a combination of skills which require the technical knowledge of graphics artists, typographers or compositors, photographers or art editors. While it may be impractical to expect to have all these resources at your fingertips or in your employ, you should not fail to realise that all these skills can be obtained from outside sources and you would do well to consider whether they should be brought in to assist the publisher of the technical work before, during or after the writing process. Failing that, and particularly if you are a 'one man band' in your organisation, responsible for all aspects of production, you should invest some time in learning some design and layout principles, even if they are of the most basic form.

DESIGN

One good source of learning about layout, is to look at as many examples as possible of contemporary technical documentation, and

DESIGN AND LAYOUT

> **TRAINING COURSE NOTES**
>
> There is no established routine within the software for security backup procedures - you are referred to the DOS backup facilities, which may be different depending upon which machine you are using, ie DCOPY, COPY or a BACKUP program.
>
> **Backing Up Files**
>
> However, to make things easier, a list of the filenames are given below.
>
> The data files are made up as follows, and all have the .DAT extension.
>
> **Data File Format**
>
> eg   A   SL   -NAME   .DAT
>   Company   Module   File Name   Extension
>   Identifier
>
> Company Identifier - is the one that you use when entering the software through the date and company identifier screen. It may be an alpha character from A to Z.
>
> Module Identifier - this may be one of the following:
>
> SL = Sales Ledger
> PL = Purchase Ledger
> NL = Nominal Ledger
> IN = Invoicing
> ST = Stock Control
> WG = Payroll (please note that WT files are also created by the payroll calculation routine)
> JC = Job Costing
> AS = Bill of Materials (Assemblies)
>
> File Name - each set of data files have their own file names. To avoid corruptions of data, all SL files, for example, have to be copied together, as the individual files are linked together with pointers. The files for each ledger are listed below:
>
> Sales Ledger
>
> **Module Files**
>
> NAME
> INDX
> PARM
> TRAN
> ANAL
> FREE
> LETT
>
> Invoicing
>
> PARM
> FREE
> SOP

Figure 4.1 Sample Layout

try to understand what effect the layout has on the overall use and appearance of the documentation. Some examples may be more comfortable to the eye than others. Try to look for reasons why. Some examples may show poor design which is detrimental to the flow of reading throughout the document, or is designed in such a way as to make the division of one section from another unclear. In software documentation, I have often come across examples of user manuals in which the graphic artist (if there was one involved) appeared to have a 'field day', with chapter headings, running heads and footers and paragraph structures designed like pieces of abstract art more suited to the cover of a sci-fi novel than technical literature.

Design has to be handled carefully. Like a powerful weapon, it can be 'dangerous' in the wrong hands. In this respect, no book such as this can be any substitute for apprenticeship. It can only serve to bring to your attention matters about which you should be aware. In addition, there are many books which specialise in composition, printing, design and typography which can be useful sources of learning in this sphere.

Generally, when planning the design of a piece of technical documentation, you will be producing finished artwork which will fall into one of two categories:

— Word processed (or typed) copy.

— Typeset or reprographic artwork.

Your design considerations are greatly increased when you begin to use the second of these processes, since you have to contend with the subtle attentions required to the form and size of typefaces, as well as the overall layout planning (this is given a little consideration later in this chapter). If, for whatever reason, you produce your artwork* using typed or word processed text, then your design and layout capabilities are more restricted than those which you can apply to the use of various typefaces, but this does not mean that you can opt out of layout planning altogether. In fact, layout becomes even more important, if you want your documentation to have more impact than a telephone directory.

*this is a general term and may represent any finished material ranging from your word processed originals from which you photocopy your documentation to the final setting which may be used as a master for platemaking for larger print runs

DESIGN AND LAYOUT

You should, by all means, use what facilities you can to enhance the text, bold headings, underlining, and so on, but these should be used wisely. Figure 4.3 shows an unwise use of such text enhancements to supposedly create some variety in the text.

Figure 4.2 shows an example of word processed text which has no thought for design included whatsoever. Unfortunately, there are many examples of such 'layouts' in existence in techncial manuals or specifications.

Figure 4.3 shows the same page of text with just too much hard work! Here the designer has gone overboard to emphasise the paragraph headings and text components to such an extent that the reader could be forgiven for feeling a little dizzy just looking at the page.

Figure 4.4, on the other hand, strikes a more acceptable balance between the two. Notice how the correct use of white space makes a significant improvement to the overall impression of clarity. The text itself looks orderly and the reader should feel more comfortable reading such a page.

Remember that these considerations, while independent of the content of the text itself, should really be considered before the text is written. If the author is clear about how the page design is going to be, then the text can be adapted to make best use of the design. For a more obvious example, if the author knows that there will be the facility to include graphics illustrations throughout the text, then this can affect the style in which the text is written, as discussed in the previous chapter. Similarly, if the page design is known in advance of authorship of the final draft, then attention can be given to the most effective way of highlighting the parts of the text which most need the reader's attention, and the organisation of the sections, paragraphs and sub-divisions of the text can be more deliberately presented. This all makes for an organised and interesting appearance to the reader.

Though you should consider the *type* of reader when you write the actual material for the document, this should not necessarily be important with regards to layout. I have often heard comments such as 'well, this documentation is for technical readers, so we needn't bother with any fancy design or anything, they just want the facts.'

7.10 Creating an Index File to Extract BAR/FEE/GDS Information

The graph specification you have just entered is to be used to create a diagram which displays information about BAR, FEE and GDS type transactions in date order. Your data file contains records of several different types. The records are not sorted in any particular order.
Before you can display the line graph, you must set-up an index which can be used to access BAR, FEE and GDS records, in date order. The procedure is similar to that explained in Tutorial 2.
- Select the Information Manager function Sort Transaction Index (ST):

```
Sorted Index - memtdate
Description - Tran recs by date
Sort field 1 - the number of the date data field (this
should be 17)
Sort field 2 - Press ENTER
Accept sort fields - y
```

The system sorts 20 records, then returns you to the Information Manager menu.
- Select the Information Manager function Specify Search parameter (SS). Follow the sequence below:

```
Operator    F (for first)
Field name  Type
Search for  = FEE
Operator    O (for Or)
Field name  Type
Search for  = BAR
Operator    O (for Or)
Field name  Type
Search for  = GDS
Operator    E (for End)
```

- Save the search parameters as follows:
```
Search file  : memfbg
Description  : finds fees/bar/gds
```

The system saves your search parameters then returns you to the Information Manager menu.
- Select the Information Manager function Search File (SF):
```
Search file - memfbg
Index name - memtdate
Option - 5:Output Transaction index
Output Index - memfbgdt
Description - fees/bar/gds by date
Disregard upper/lower case - y
```

The system locates and sorts 12 records, then returns to the Information Manager menu when you press space. You should recreate the index `memfbgdt`, by following the above

Figure 4.2 Sample Word Processed Text

DESIGN AND LAYOUT

7.10 Creating an Index File to Extract BAR/FEE/GDS Information

a) The graph specification you have just entered is to be used to create a diagram which displays information about BAR, FEE and GDS type transactions in date order. Your data file contains records of several different types. The records are not sorted in any particular order.

b) Before you can display the line graph, you must set-up an index which can be used to access BAR, FEE and GDS records, in date order. The procedure is similar to that explained in Tutorial 2.

   -____Select_the__Information__Manager_function__Sort Transaction_Index_(ST):

Sorted_Index - **memtdate**
Description - **Tran recs by date**
Sort__field_1 - **the number of the date data field (this should be 17)**
Sort_field_2 - **Press ENTER**
Accept_sort_fields - **y**

c) The system sorts 20 records, then returns you to the Information Manager menu.

   -___Select_the_Information__Manager_function_Specify Search_parameter_(SS)._Follow_the_sequence_below:

Operator   **F (for first)**
Field_name **Type**
Search_for **= FEE**
Operator   **O (for Or)**
Field_name **Type**
Search_for **= BAR**
Operator   **O (for Or)**
Field_name **Type**
Search_for **= GDS**
Operator   **E (for End)**

   -__Save_the_search_parameters_as_follows:

Search_file   : **memfbg**
Description   : **finds fees/bar/gds**

d) The system saves your search parameters then returns you to the Information Manager menu.

   -___Select_the__Information_Manager__function_Search File_(SF):

Search_file - **memfbg**
Index_name - **memtdate**

Figure 4.3   Over-presented Layout

7.10 Creating an Index File to Extract BAR/FEE/GDS Information

The graph specification you have just entered is to be used to create a diagram which displays information about BAR, FEE and GDS type transactions in date order. Your data file contains records of several different types. The records are not sorted in any particular order.

Before you can display the line graph, you must set-up an index which can be used to access BAR, FEE and GDS records, in date order. The procedure is similar to that explained in Tutorial 2.

a) Select the Information Manager function Sort Transaction Index (ST):

    Sorted Index   memtdate
    Description   Tran recs by date
    Sort field 1   the number of the date data
                              field (this should be 17)
        Sort field 2   Press ENTER
        Accept sort fields   y

The system sorts 20 records, then returns you to the Information Manager menu.

b) Select the Information Manager function Specify Search parameter (SS). Follow the sequence below:

    Operator   F (for first)
    Field name   Type
    Search for   = FEE
    Operator   O (for Or)
    Field name   Type
    Search for   = BAR
    Operator   O (for Or)
    Field name   Type
    Search for   = GDS
    Operator   E (for End)

c) Save the search parameters as follows:

    Search file   memfbg
    Description   finds fees/bar/gds

The system saves your search parameters then returns you to the Information Manager menu.

d) Select the Information Manager function Search File (SF):

Figure 4.4 Well Presented Word Processed Text

DESIGN AND LAYOUT

Figure 4.5   Alternative Arrangements for Text Layout on a Double Page

This misconception, that a technical audience prefers to read boring lists of technical facts through pages of photocopied word processed text, leads to a lazy approach to the documentation.

All readers, whatever their level of interest in the material, should be given the benefit of good presentation by careful design. This applies as much to a technical fault report of a development process as it does to a user manual for a technical product. For those of you who have any amount of technical documentation produced by word processor, you will have a certain amount of copy manipulation facilities at your fingertips with most word processor packages. Setting tabs, margins, page lengths and column widths, which in effect determine the layout of the page, becomes part of the layout process. Given that all well known word processors have such facilities for setting formats, then each has some in-built layout and make-up facility (see also Chapter 6 on word processors).

PLANNING THE TYPE AREA ON THE PAGE

Try to imagine your type or text areas as blocks on the overall page. A variety of effects can be achieved by arranging blocks to create the overall space usage requirements on the page itself. The increasing of white space surrounding the text areas can serve many purposes, for example:

— provide space for notes;
— create an impression of space which makes the text look like lighter reading;
— gives distinction to the text.

The reduction of white space around the text areas can serve to do the following:

— enable more text to be printed on the pages, hence fewer pages in the document;
— make the text appear crowded if reduced too much;
— improve the appearance if the type itself is openly spaced or has leading of above average values.

Text Columns

Look at the examples in Figure 4.5. The arrangements of the text

DESIGN AND LAYOUT

blocks can be considered to create a balance between type and space that best suits the purpose of the documentation and the aesthetic qualities of the appearance. If you choose a multiple column format for word processed output, do not make the columns too narrow. Two columns on an A4 page is quite adequate, as three columns will begin to appear messy in standard courier typewriter typefaces of 10 or 12 point.

If you decide on a single block of text, how do you best place it on the page in respect of the paper size? Generally, most people go for placing it in the middle, and this will usually suffice so long as it is placed in the 'optical' middle, and not the 'geometrical' middle.

If you place the text on the paper such that the white space above and below the block of text is equal by measurement, the text will 'droop', appearing to lie beneath the middle, (see Figure 4.6). In order to make amends, you have to make an optical adjustment to move the block up until it appears correct according to your eye.

In layout, you will probably find that eye judgement is better than relying on strict measurement with a ruler, since no reader will get out a ruler to the page, and the impression formed of the layout, even though it may be purely a subconscious impression, will be a product of his or her feelings.

You will have to rely upon experience to make the necessary judgements about optical alignment and I for one have learned more through correcting and improving previous work by conscious effort than can be taught through a strict set of guidelines or tutorials on the subject. Using wide margins can give text an appearance of distinction, but may be a totally impractical luxury for an essentially technical communication.

So, layout must be practical as well as simply aesthetic. If you get yourself a book on design and layout, you may find little application for much of the information contained within, especially when it comes to sections that deal with making compensations for 'visual tensions', 'symmetry' and 'asymmetry', and 'the balances of stresses and reliefs'. This is strictly for the advertising, magazine and newsprint designers.

You will do better to learn from your own experimentation. By

12 Enquiries

Select function 5 (Enquiries) from the payroll processing menu.

The screen will display a box in which you should enter the employee number to whom you wish to enquire, and press enter.

You will be asked if you require a hard copy (printed copy) of the enquiry. Reply 'Y' or 'N'.

The printout may be directed to the printer, to the screen or to a disk spool file (although on this occasion there is no value whatever in directing the printout to the screen). Press the enter key if you wish to change the destination of the printout. Each time you press the enter key, you will see the selected destination of the printout on the bottom line of the screen.

If the printout is to be directed to the printer, ensure that it is loaded with plain listing paper, that it is aligned to the top of the page, and the printer is switched on-line.

Press the space bar to continue.

The enquiry will be printed on the screen as well as the printer and will show the following details:

a) Employee name and number
b) Analysis code
c) Tax code and N.I. number
d) Payment method
e) Cost centre
f) Brith date
g) Leave date
h) Days/week
i) Sex
j) Leaver indicator
k) On holiday indicator
l) Bank details and autopay code
m) Gross pay, tax and net pay during the last pay period
n) Gross pay, tax and pension contributions for year to date
o) Statutory sick pay paid to date
p) **Gross pay and tax in previous employment**

Figure 4.6 Text Seems to 'Droop' When Centred Vertically

DESIGN AND LAYOUT

OSPREY INVOICING

a) Company Name

Enter the company name as you would wish it to appear on your invoices and report headings.

b) Four Address Lines

Enter the address of your company as you would wish it to appear on your invoices.

c) The next four lines are nominally for two telephone numbers, a telex number and Employer I.D. number. However, they may be employed in any way you wish. For example, any or all of these entries could be used to show payment terms. These details are printed on invoices only if you request them and are for no other purpose. It is useful when using these four parameters to put a note item in, for example 'TEL:', 'TELEX:', etc.

d) Printer Type

Enter the code for the type of printer you propose to use. These codes are summarized in section 13 of the General Reference Guide.

e) Top of the Form Code (TOF)

This is the code that instructs the printer to move to the head of the next sheet of paper. The standard ASCII code for all modern printers is 12. However if your printer does not have the facility to automatically feed to the top of the next form, but instead performs line feeds, you should **enter 0** (zero). If you are in any doubt, refer to your supplier.

f) Printer Columns

This is the width of the paper (in columns) used in your printer, and will either be 80 or 132 depending upon the printer. 80 column printers will use a 2 line format for certain reports. 132 column printers will only use 1 line, the notational items being attached to the extreme right.

Figure 4.7   Open Paragraph Setting

INCOMPLETE RECORDS

13 Errors and general information

13.1 As far as is practical, errors are handled by the computer. Checks are applied to data as it is entered from the keyboard and data which logically could not be correct is not accepted. Instead, an error message is displayed and the program permits you to re-enter the data.

If errors occur in opening files or in reading or writing data, an error message is displayed. Step by step instructions are then given on the screen, indicating the action that should be taken.

Nevertheless, in some error conditions further explanation may help and this is provided below.

13.2 CORRECTING AN ENTRY ON THE SCREEN
The delete key may be used to erase the field. Any character may be overwritten.

Delete the entry by tagging it with d and make the correct entry. Entries cannot be amended once they have been completed.

13.3 CORRECTING AN ENTRY IN THE COMPUTER
Delete the entry by tagging it with d and make the correct entry. Entries cannot be amended once they have been completed.

13.4 ERRORS ON OPENING FILES AND FILES NOT FOUND ERRORS
The most common cause is calling for comparative figures when you have not set up the file but in this case the computer will continue its operation after displaying the message. Other than this, the wrong disk is probably loaded or you may not have reloaded a disk as instructed on the screen.

13.5 OPENING ENTRIES ALREADY SET UP
This will occur if you try and set up comparative figures twice. You are offered the option of continuing and overwriting the previous year's figures on the disk, or aborting.

13.6 DISK ERRORS — 57 OR 70, 71 OR 72
The errors occur when the computer is unable to read from or write to the disk. The most common causes are either: you are trying to write to a disk protected with a write-protect tab or the disk is not properly home in the drive.

If you check these and continue, the program will retry.
. If the error message re-occurs then you probably have a bad disk and you will have to press Esc to exist and retry with a back-up disk if reading.

If writing then the data is lost and will have to be re-entered.

Figure 4.8 Closed Paragraph Setting

DESIGN AND LAYOUT

simply making a conscious effort to consider layout, especially if you haven't really given the subject a great deal of thought before, you will surprise yourself with how much improvement you can make on your own to the techncial documentation's presentation. Apart from that, it can be a satisfying diversion from worrying about the purely technical content of the document.

A word of warning though; you can get a little too carried away with layout experiments. You will sometimes find that unbalanced or unusual layouts can be quite visually effective, but you should not use such designs universally. This is because unusual layouts tend to be rather fatiguing over a large document, especially manuals, and weaker than layouts which tend to adhere to a more traditional format. If your page layout is anything but simple, your reader may become distracted by this.

Paragraph Spacing

Using paragraph spacing liberally can give the text an open feeling and allows 'breathing space' for the reader. Figure 4.7 shows a page from a Pegasus Software manual, typeset with an open paragraph style. The appearance to the reader is one of a 'clean' image which does not seem to make the same demands on the reader's brain as, say, a more solid text style, as shown in Figure 4.8.

CHAPTER BEGINNINGS AND HEADINGS

Use chapter beginnings to relieve the uniformity of the main body of the text. In some technical documentation I have seen, it is not clear where one section begins and another ends, simply because the new chapter headings are too confused with the rest of the text, or sub-headings are given equal weighting to main headings.

If you are word processing your documentation, and therefore do not have the typographic freedom of typesetting at your disposal, then you will need to be even more stringent about making new section headings and chapter beginnings stand out.

FOOTNOTES AND RUNNING HEADS

If your artwork is based on word processed or typed copy, you have limited alternatives about the placing and size of footnotes, folios and

Running Head

1.2 The Text Heading

The size of the running header in relation to the text heading is such that it seems to have more importance that the heading within the body of the text itself. Using bold rules can be distracting to the eye, rather than enhancing the page layout.

Running Head

1.2 The Text Heading

In this example, however, the strength of the running header is reduced so as not to disrupt the page design and distract the eye. The importance is placed on the text heading which draws your attention to the following paragraphs of text. The weight of the rule beneath the running header is sufficient to provide an interesting design, but thin enough not to attract too much attention.

Figure 4.9 Weighting of Running Heads

DESIGN AND LAYOUT

System Builder System-File-Field Definition

Refer to section 2.2, paragraph 'File Types' for details on the differences between the file types if you are not sure what to choose.

Record Length (New Files Only)
The appropriate record length for the file selected is displayed and subsequently increments as fields are added to the file definition. The start record length depends upon the file selected:

```
CSAM           = 1
EXTENSION      = 1
RANDOM         = 1
TRANSACTION    = 5
SEQUENTIAL     = 0
PEG.EXTENSION  = 0
```

Once the record length is displayed, the prompt shows:

'Proceed to create new file (Y/N) Y ',

and responding Y will instruct the system to create the file on disk that you intend to define fields for in the following part of this entry.

46

Figure 4.10 Use of Rules as Page Footers

6.5 POSITIONING FIELDS ON YOUR REPORT

Defining a report is very similar to defining a screen. The maximum size of a report definition is 254 columns by 100 lines and your 80 by 23 screen is used as a window on your report layout. By moving the cursor off the screen, (left, right, up or down) the screen will jump across to the next report section.

The keywords @HEADER, @FOOTER, @BODY, @TOTAL, are typed onto the report format itself. They must start in column 1 and they must be typed as either all capitals or all lower case, or they will not be recognised. No other text or fields may appear on the same line as the above keywords.

Report loops are defined like page displays, with a backslash (\) in column one.

Field information is handled in the same way as for screen, except that in reports, there is no input process. Files can be referenced automatically. If the program comes across a field defined from a file which has not yet been read, it will try and read that file with its current key value (if automatic referencing is on).

When a numeric field is defined using CTRL+F, you will be given the opportunity to total it (see below) into an element of the array REPTOT. More than one field can be totalled into the same element of REPTOT.

Up to 10 files can be accessed in one report.

Layout Commands
To display a list of the options available through the use of function and control keys, type ALT-H and Help information is provided:

Alternatively, use your 'Quick Reference Guide Card' for help with report definition functions.

116

Figure 4.11   Using Rules to Divide Sections

DESIGN AND LAYOUT

running heads. Basically, you either place them centrally, or right or left align them above or below the text blocks according to your taste. If you typeset the text, then generally you should choose to set footnotes and running heads in a smaller type size. Figure 4.9 shows examples of the weighting of running heads and shows how you can overdo their importance in relation to the rest of the text.

Use ruling separators with discretion. Do not make them too heavy as they appear to dominate the page. Using a thin rule at the foot of a page can be optically useful on pages where the text area often falls short of the footer. The line 'draws' the eye to the foot of the page and makes it appear more complete optically (see Figure 4.10).

Rules may be useful for opening chapters or sections too, as shown in the example illustrated in Figure 4.11.

SOME FUNDAMENTALS OF TYPE

Without giving a long-winded history of typography, for those who are new to the subject, in 1440, Johann Gutenberg (from Mainz) invented a process of casting separate letters in matrices struck from punches, and the setting of type with these letters. Before Gutenberg, all text was written by hand or reproduced from a block or woodcut (an engraved board). Since Gutenberg, principles have changed little, only the technology with which to do it; but it brought about the introduction throughout the ages of the design of varied type styles, and handwritten letters were adapted and developed into new typefaces.

Roman forms of type were among the earliest developments into typefaces. At the time of Gutenberg there was no uniform system of measurement for type, nor any standardisation for the size and height of typefaces. It was not until Firmin Didot (1764-1836), a Frenchman, introduced a method called the 'point' system that type sizes became more standardised.

Anyone concerned with type and printing in any form should become familiar with the point system. The point is the basic unit of measurement in typography, and all terms used in printing derive from this one measurement. In North America and Great Britain, the point is approximately 1/72 of an inch (.351mm) and is called the pica point. In Europe, the point is a little larger (.376mm) and called the Didot point. In both cases, points are used to describe the length of one metal chunk of type (see Figure 4.12).

The metal letters cast for early type processing systems were set in the composition matrix by a hand-setter, from left to right and upside down with each line adjoining the previous one. The 'nick' enabled a typesetter to place the letters in the correct position without having to look at them, and nicks in different forms and heights differentiated types and variants of the typeface from one another.

A 'pica' is 1/6 of an inch, or 12 points and is used to measure the width of columns and pages, for example. You also need to understand the difference between the actual height of the character impression on a metal cast character compared with the point size of the type. To illustrate, a letter M in 72 point on metal type is a character which is cast onto the top of the metal block. The block's surface is exactly 72 points (one inch) in height, but the actual letter M will be smaller than the overall size of the metal, so, by tradition, the point size of a type refers to the dimension of the metal, and not the character. The difference is necessary because the space is required for 'ascenders' and 'descenders' (see Figure 4.13).

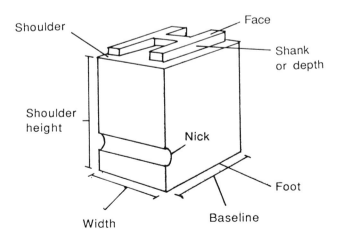

Figure 4.12 One Character Type

DESIGN AND LAYOUT   67

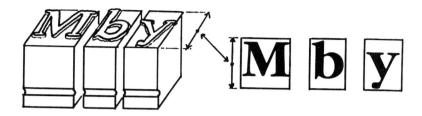

Figure 4.13   Metal Type Chunk and View of Typeface Show Space Required for Ascenders and Descenders

Composing also required the use of materials for spacing and separating letters or words. Leads (pronounced leds) established the distance between lines. Where text lines were separated from one another only by the point size of the type (without leads or leading), this was called setting solid (see Figure 4.14).

Setting solid can be perfectly acceptable in some types, but is dependent upon the characteristics of individual typefaces. If an unsuitable typeface is set solid, then ascenders and descenders are liable to touch, resulting in the kind of collision of characters illustrated in Figure 4.14.

Where text was given more breathing space, which could help make it more legible, thin strips of lead were inserted between the rows of metal characters, and gave rise to the terminology of leading. In such cases, one refers to the point size of the face being set on a body size which allows this leading and is referred to in the manner, for example, 10 on 12 (10 point type on 12 point body, giving 2 point leading). Thus 10 on 10 or 12 on 12, etc is setting solid.

When applied to the fundamentals of design and layout, the choice of typeface and size plays an important part in creating the right kind of overall image. The differences may be subtle, but the effect can

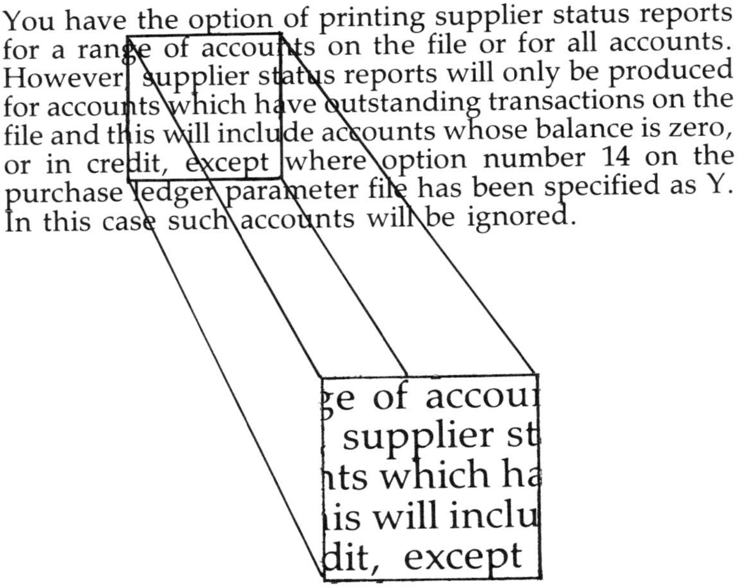

You have the option of printing supplier status reports for a range of accounts on the file or for all accounts. However, supplier status reports will only be produced for accounts which have outstanding transactions on the file and this will include accounts whose balance is zero, or in credit, except where option number 14 on the purchase ledger parameter file has been specified as Y. In this case such accounts will be ignored.

Setting solid can cause collision of characters

You have the option of printing supplier status reports for a range of accounts on the file or for all accounts. However, supplier status reports will only be produced for accounts which have outstanding transactions on the file and this will include accounts whose balance is zero, or in credit, except where option number 14 on the purchase ledger parameter file has been specified as Y. In this case such accounts will be ignored.

Setting with one point extra leading

Figure 4.14   Setting Solid and One Point Extra Leading

DESIGN AND LAYOUT 69

> Pegasus *Multi-User* Introduction
>
> ## 4 COMPANY PARAMETER UPDATE ROUTINE
>
> These 'parameters' will store the details of your company name and address, home currency, printer types etc. They are common parameters for each of the *Pegasus M-U* modules (i.e. Sales, Purchase, Nominal etc.) for one particular company identifier. After each entry typed in the parameters, **press the enter or return key**.
>
> ```
> PEGASUS                    Company Parameter Update              23 Jun 87
> 
> Company Name                    :SEA - SURE PRODUCTS (M-U) LTD
>         Address Line 1          :Brikat House
>         Address Line 2          :35-41 Montagu Street
>         Address Line 3          :Kettering
>         Address Line 4          :Northants
>         1st Telephone No.       :Tel.(0536)522822:
>         2nd Telephone No.       :
>         Telex No.               :Telex 341297
>         VAT Reg No.             :VATno313 1976 17:
> Abbreviated Company Name        :SEA - SURE:
> Master Password                 :*
> Access Password                 :sb
> Number of Decimal Places        :2:
> Home Currency Description       :STERLING
> Tax Description                 :VAT:
> Is a Log file required          :Y:
> 
>                         Enter Y or N for Log file
> ```
>
> ### 4.1 Company Name
>
> To a maximum of 40 characters in length, **enter your company name** here, as you would wish it to appear on such printouts, for example, as statements, debtors letters, remittance advices etc.
>
> ### 4.2 Company Address Lines 1-4
>
> Each line of your company address may be entered up to 25 characters in length.
>
> ### 4.3 1st and 2nd Telephone Numbers
>
> For those reports mentioned in paragraph 4.1, and others, such as invoices, order acknowledgements etc., it may be useful to include your company's telephone number details. **Enter a telephone number up to 16 characters long**.
>
> 10   Issue 3 — May 1987

Figure 4.15   Sample Page Showing Poor Choice of Type Size and Leading

Pegasus *Senior* Introduction

4 Company Parameter Update Routine

These parameters will store the details of your company name and address, home currency, printer types etc. They are common parameters for each of the Pegasus *Senior* modules (i.e. Sales, Purchase, Nominal, etc.) for one particular company identifier. After each entry typed in the parameters, press the enter or return key

4.1 Company Name

To a maximum of 40 characters in length, enter your company name here, as you would wish it to appear on such printouts as, for example, statements, debtors letters, remittance advices, etc.

4.2 Company Address Lines 1-4

Each line of your company address may be entered up to 25 characters in length.

4.3 1st and 2nd Telephone Numbers

For those reports mentioned in paragraph 4.1 and others, such as invoices, order acknowledgements, etc., it may be useful to include your company's telephone number details. Enter a telephone number up to 16 characters long

9

Figure 4.16 Same Page Set in Smaller Typesize

DESIGN AND LAYOUT

be quite noticeable if you compare one with another. Take a look at the page from a user manual illustrated in Figure 4.15 (courtesy of Pegasus Software Limited again).

It may not be apparent that there is anything wrong with the chosen typeface or type size when observed in isolation, but a change in typesize and leading, even retaining the same typeface (Garamond in this case) creates quite a subtle, yet effective difference to the overall appearance of the page, as shown in Figure 4.16.

Considering this is only one element of page design and layout, you should begin to appreciate the tremendous impact that can be effected upon a page by correct use of the design elements.

If you choose to use multiple columns on a page, then consider how many columns you are going to use and why. The optimal reading width of a line is about 50 to 55 letters, which corresponds to a line 12 or 13 picas long for 6 point type, 15 or 16 picas for 8 point type, 17 or 18 picas for 9 point, and 19 or 20 for 10 point. With 12 point and larger, lines of 23 to 30 picas are often used in books or manuals. The example page of a Pegasus manual, shown in Figure 4.16 was set in Garamond as 10 on 11, with a measure of 26 picas.

CONCLUDING DESIGN AND LAYOUT

Layout and design can be fun. Most importantly, they should not be ignored if you wish to give a good impression to the reader. However much emphasis one places on the technical accuracy of the content of technical documentation, you should not forget that its primary function is to convey information. That means that someone has got to read it. The more appetising the document is to read, the more chance the reader will feel comfortable using the documentation. If layout is bad, you may even dissuade the reader from going near the document at all.

You will need to experiment to get the best results, and how you go about it will depend very heavily upon the facilities you have to hand to do the task. A word of advice for users of desktop publishing packages; if you think you can successfully design useful layouts on screen as you go along, with or without the aid of a mouse, then you are kidding yourself. Until you have acquired the skills that can only be gained by experience in layout and design, your best bet is to stick to a sheet of graph paper or use a layout pad (available from most office stationery suppliers) and a pencil and ruler. When you have created a few rough layouts to choose from, you can choose the best before even contemplating making up layout format files on some software package. Don't just play with the technology because it is there, and certainly don't consider the documentation supplied with it as the best tutorial for the subject.

Finally, layout can be enhanced by the use of graphics and illustrations, and colour. I have deliberately avoided discussing the use of colour, since this is not only subjective, as is much of layout principles, but severely affects the cost of production. If you have the financial resources to apply colour to your documents, then you will almost certainly improve their image if the colour is used tastefully. Remember that every colour you want to add to your final document involves a further print process as the paper goes through the press.

Colour separation has to be checked and the proofing cycles become inevitably more time consuming and therefore expensive. The use of a second or third colour can be useful, however, to add a touch of interest to the overall appearance of the document, as well as providing a further vehicle for highlighting text, headings, chapters, illustrations, etc. As for illustrations and graphics, these are considered in the following chapter.

5 Illustration

All courses and instruction books on technical writing will emphasise the benefits of illustrations in your documentation. A picture's worth a thousand words, as Confucius said. The possibilities for illustrating documentation have become significantly widened with the advent of desktop publishing technology. With high quality output devices like plotters, graphics printers and laser printers, the need for hand-drawn technical illustrations is dwindling, and the modern technical illustrator is having to become adept at using this new technology in order to remain competitive with the increasing production of illustrations in-house.

A good line diagram can be useful in amplifying and clarifying main points in the text. Apart from its obvious benefit in this respect, the use of illustration material in a technical document, whether line illustration work or photographs, helps to break up the text, which might otherwise seem monotonous in a large document, and thereby retain the reader's interest.

Just how much illustration work to use will depend on a number of factors, which include what facilities you have for producing illustrations, and how much costs are increased by their inclusion in the text. These two factors may be closely linked. If you have in-house facilities for reproducing line artwork through some graphics device, then you may be able to produce camera-ready artwork without involving the costs that would otherwise be incurred if you had to employ the services of technical illustrators or graphic artists.

Of course, where illustrations are essential, for example in parts diagrams, circuit diagrams, flow control charts, etc, you may already be facing considerable costs in documentation production in any case. If you haven't already done so, you would do well to investigate the current technology for computer graphics devices, and calculate the cost benefits of using this technology against the more traditional approaches.

Many of the larger companies, with significant documentation departments, may employ the use of specialist CAD/CAM systems for producing illustrations based upon those used for the design and manufacture of the items for which the documentation is associated. For more modest in-house installations, the facilities of drawing packages for PCs are nonetheless quite useful tools for enhancing the quality of illustration material, and many software packages enable the graphics of diagrams, charts, and even scanned photographs to be integrated with the text of the document itself for simultaneous output.

It is worth looking briefly at some of the new technology facilities that are available for assisting in illustration work, and then discussing some of the reproduction processes that have to be considered.

SCANNERS

Image scanners are one of the more recent desktop devices that have come to the aid of the documentation illustrator. They are not restricted to graphics only, however, but some scanners are better at this than text, although most handle both. The principle of the scanner is not far removed from that other piece of now familiar office automation — the digital fax machine.

Basically, when an image is fed into the scanner, it passes over a device called a charged coupled device (CCD) which shines light onto the surface of the image. The reflected light is captured by light sensitive cells that can emit a digital signal which is dependent upon the intensity of the light received. These digital signals are then sent to the receiving computer, which converts the signals into the necessary information required by the computer to reconstruct a 'bit image' of the original as a series of black and white areas. The black areas are reproduced as dots on the screen or output device of the computer which is receiving the image, and this process is known

ILLUSTRATION

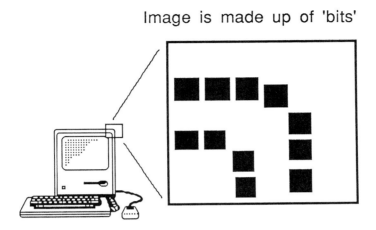

Figure 5.1 Bit-mapping

as 'bit-mapping' (see Figure 5.1). The smaller the dots, and the more of them there are to a given area, the more accurate will be the reproduction of the image.

Some scanners are quite good at producing a sharp image of a line diagram which may or may not include some text. You should realise, however, that at the time of printing (1988), unless you are going to spend a good deal of money, or wait for better quality of technology, there are not many desktop scanners that are capable of reproducing a black and white photograph with real accuracy. This is largely due to the fact that scanners are not generally very good at recognising scales or shades of grey. This is changing, as there are already scanners capable of handling up to 64 levels of grey scale, compared to the more common 16 or 32 levels. Of course, with many such devices, you get what you pay for, so the better the resolution quality and speed of processing an image, the higher will be the price of the scanner.

Scanners are well worth looking into if the amount of illustration work you need is quite high and especially if much of it is already drawn or available and simply needs integrating with the text in a document stored on a text processing device. There are scanners

76 THE TECHNICAL DOCUMENTATION HANDBOOK

which can handle reproduction of pages from books, working rather like a photocopier, and these are the 'flat bed' plotters. Take care that you watch the copyright of illustrations which you may 'pirate' for your documentation, however.

GRAPHICS/DRAWING SOFTWARE

Software for graphics has gone through quite a boom in the past couple of years. This has been largely due to the impact made with the introduction of desktop publishing. DTP software has demanded high resolution screens and output devices which has opened the way for more and more sophisticated graphics packages. The software that handles graphics ranges from simple drawing tools that enable a line diagram, flow chart, or something similar to be produced on a screen and output to a dot matrix graphics printer, to mouse-driven illustration software for producing high resolution artwork through laser printers.

There is a great deal to choose from in-between, and much will depend upon the illustrator's general needs. Graphics packages have

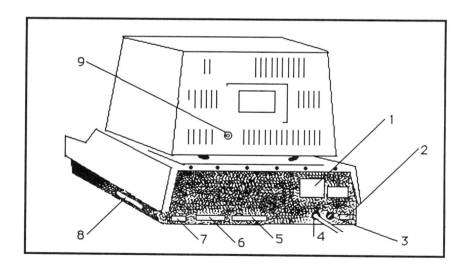

Figure 5.2 Example of Line Drawing from Electronic Graphic Device

ILLUSTRATION

been around for quite a while and plotters have been in use within the electronics industry for printing complex circuit diagrams that have been mapped on a computer imaging device. The major step forward in illustrating has been that previously specialised tools are now readily available on the desktop PC. Prices vary considerably too.

Additionally, word processors are becoming more sophisticated, and many of them now provide an element of drawing or graphics facilities. There are more details in the chapters on word processing and desktop publishing.

CAD/CAM SYSTEMS

As with graphics software in general, CAD/CAM (Computer Aided Design and Manufacture) Systems have been previously confined to special dedicated computer graphics systems, and have been, and still are, in use in the more specialised or larger publications environments for producing high quality illustrations. Now this technology is also reaching the PC environment, making the tools readily available to the more modest area where illustration work is probably engineering based.

LINE COPY/DRAWINGS

Line copy is any image that is made up of solid black, with no gradation of tone. The text of this book is also line copy, but the term line artwork is more commonly associated with line drawings used to illustrate a text, and are reproduced from hand drawings or more recently from electronic graphics devices (see example in Figure 5.2).

Whether the drawing is by hand or from a graphics device, you should consider the final size of the illustration in the document. If you are including the image as integrated graphics in, say, a desktop publishing software package, then you will probably scale the illustration automatically to fit whatever space you have assigned for it on the page.

If you are using a more traditional approach, where an illustration is to be pasted into the artwork of the document at a later stage, then you should ensure that the illustration is at least as large as its final size.

If the illustration is smaller than it will eventually be printed, then the enlargement process necessary to scale up the illustration will affect the clarity. In making enlargements, lines generally become ragged and unclear.

It is a good idea to have line drawings reproduced through a camera process into a bromide (a bromide is a black and white image, which may be text and/or line work reproduced onto high-contrast photographic paper). During the camera process itself, the illustration may be scaled to the final artwork size. It is not unusual, therefore, for the originals of line illustrations to be made two or three times larger than the final version, so that the lines will appear prominently and clearly in the reproduction. Be careful, however, not to reduce illustrations to less than 25% of the original, since they will often turn out to be unattractively compressed.

HALFTONES

Photographs will not reproduce with any clarity if you use an original photograph as final artwork. You have probably noticed how a black and white photograph with continuous shades of grey, does not reproduce very well on a photocopier. In order to make reproduction of photographs in printed documentation appear with some degree of clarity, the original image must be broken down into patterns of dots or lines, through a process known as screening.

There are two kinds of screening techniques, one which uses glass screens and one which uses contact screens. A glass screen consists of a very finely ruled grid pattern. The screen is placed between the original photograph and a camera which takes a copy of the original. A contact screen is a pattern of vignetted dots which is placed in direct contact with the unexposed film of the camera making the copy of the original. Both processes produce the same result, which is to reduce the original continuous-tone image into thousands of tiny dots, which vary in size, shape and number. When reproduced, the dots form the illusion of the original shades or tones. Screens are measured by the number of lines per inch, and this can vary from 55 to 150. The more lines per inch the screen has, the better the quality of the reproduction, or the higher the resolution. To illustrate, a 55 line per inch screen will have approximately 3,000 dots per square inch, whereas a 150 line per inch screen will have in the order of 225,000

ILLUSTRATION

Figure 5.3 Example of High Resolution Screened Photograph

dots per square inch. Therefore, in order to reproduce the original in the most accurate representation, a halftone screen of high resolution is required (see Figure 5.3).

The amount of detail you do decide on cannot be simply a matter of personal choice. One consideration is the quality of the paper on which the halftone will be printed. You may have otherwise wondered why all halftone screens are not 150 lines per inch in order to produce the maximum quality reproduction. The answer lies in the need to suit the resolution of the screen to the surface of the paper on which it is to be printed. For example, a newspaper uses a rather rough paper for the sake of economy, inks of a watery consistency and presses not especially designed for fine reproduction. If a fine screen were used, the paper would be unable to hold the detail and the spaces between the dots would fill in. Consequently a newspaper will use screens of about 85 lines per inch. By contrast, the type of paper used in magazines, by its nature being smoother, is capable of reproducing a finer screen of up to 150 lines per inch.

COMBINING LINE WITH HALFTONE

In some cases, an illustration may require the combination of a halftone, say from a photograph of a product, with some line work included for text, lines for pointers to components, etc. In such cases, it is not wise to include the line work on the photograph before screening. If the line work or text is screened, then it will be printed with a slightly grey appearance due to the fact that it has been broken up into dots like the continuous-tone copy. Instead, you should arrange for the continuous-tone and the line artwork to be shot separately, with the two combined together at the time of reproduction. In this way, the line work will retain its intensity of the black (see Figure 5.4).

RETOUCHING

Using the screening process enables the retouching of photographic artwork with some precision. In some cases where strong blacks and whites are broken into dots, photographic originals may have to be retouched. They may then look artificial to the naked eye, but will reproduce to good advantage. If reproduction photographs show defects, these will be more noticeable in the final halftone. Photographs of hardware, for example, can look unattractive in black

**THIS TEXT IS
TREATED AS
LINE ARTWORK**

**THIS TEXT HAS
BEEN SCREENED
AS WOULD A HALFTONE**

**Figure 5.4 Example Showing Screened
and Non-screened Text**

ILLUSTRATION

and white unless retouched, or professionally photographed in a studio with careful lighting. An experienced retouch artist will know how to produce tones of grey and avoid very delicate tones in a photograph.

SCREEN ILLUSTRATIONS FOR SOFTWARE DOCUMENTATION

If you are involved in reproducing illustration artwork for software documentation, you will probably want to include (if you don't already) illustrations of the screens that appear on the computer during the processes which are being described within the text of the documentation. Many examples of existing software documentation show that the screen illustrations are often made up by a manual process of typesetting or illustration work, without using the facilities of the computer for reproducing screen artwork itself. If you mock-up the screens you will probably find that it is a tedious process. Typesetting the screens within a box to represent the screen outline is a common way of representing screens, and this is further enhanced by the use of a different typeface, to represent that it is a screen illustration rather than being part of the rest of the document's text.

If you have the programming ability in-house, it may well be worthwhile writing a simple routine to dump the graphics of a screen onto a graphics printer to provide the screen illustration. There are several ways of solving the screen illustration problem. I know of one author who uses the print screen facility provided on his PC to produce the screen artwork. This is then fed into a scanner and scaled into the text on the document held within Ventura Publisher software, which he uses to output to a laser printer for final artwork.

Another example is the use of a software package that captures a screen image and saves it to disk as an image file that can be read into a variety of graphics software packages or DTP packages for output.

If you use PCs (IBM PCs or compatibles) there is a useful screen print software package that reproduces, in high resolution, an A4 representation of any screen displayed on the PC. The utility software is loaded into background memory and can be activated by a combination keystroke at any time, even whilst other software is loaded. This has the advantage of allowing the author to load the software about which he or she is writing and, at the appropriate

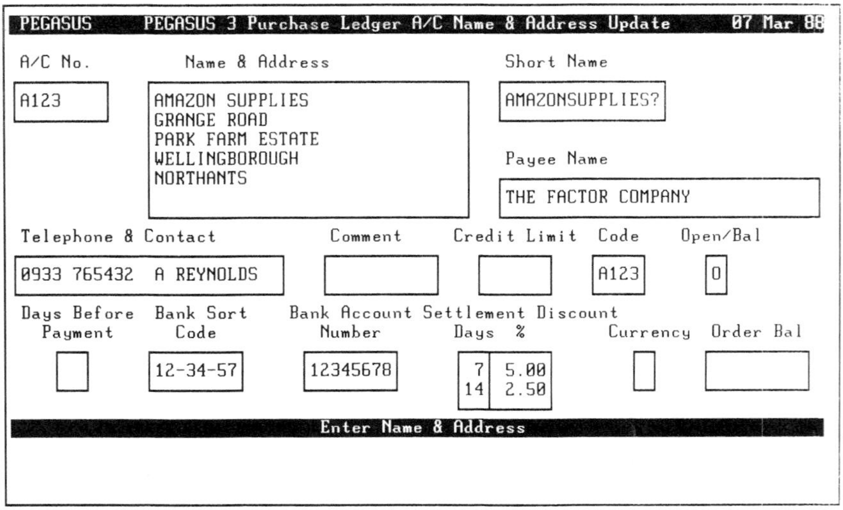

82   THE TECHNICAL DOCUMENTATION HANDBOOK

```
┌─────────────────────────────────────────────────────────────────┐
│ PEGASUS    PEGASUS 3 Purchase Ledger A/C Name & Address Update   07 Mar 88 │
│  A/C No.      Name & Address              Short Name             │
│ ┌─────┐  ┌──────────────────┐           ┌──────────────────┐     │
│ │A123 │  │AMAZON SUPPLIES   │           │AMAZONSUPPLIES?   │     │
│ └─────┘  │GRANGE ROAD       │           └──────────────────┘     │
│          │PARK FARM ESTATE  │                                    │
│          │WELLINGBOROUGH    │            Payee Name              │
│          │NORTHANTS         │           ┌──────────────────┐     │
│          └──────────────────┘           │THE FACTOR COMPANY│     │
│                                         └──────────────────┘     │
│  Telephone & Contact         Comment    Credit Limit  Code   Open/Bal │
│ ┌──────────┬──────────┐    ┌────────┐  ┌────────┐  ┌─────┐ ┌──┐  │
│ │0933 765432│A REYNOLDS│    │        │  │        │  │A123 │ │0 │  │
│ └──────────┴──────────┘    └────────┘  └────────┘  └─────┘ └──┘  │
│  Days Before  Bank Sort    Bank Account  Settlement Discount     │
│    Payment      Code         Number       Days   %   Currency  Order Bal │
│    ┌──┐      ┌────────┐    ┌────────┐   ┌──┬─────┐   ┌──┐  ┌──────┐ │
│    │  │      │12-34-57│    │12345678│   │ 7│5.00 │   │  │  │      │ │
│    └──┘      └────────┘    └────────┘   │14│2.50 │   └──┘  └──────┘ │
│                                         └──┴─────┘                │
│ ▓▓▓▓▓▓▓▓▓▓▓▓▓▓▓▓▓ Enter Name & Address ▓▓▓▓▓▓▓▓▓▓▓▓▓▓▓▓▓▓▓▓▓▓▓▓ │
│                                                                  │
└──────────────────────────────────────────────────────────────────┘
```

Figure 5.5   Sample Screen Dump Output

screen, issue a command to dump the screen contents to an Epson type graphics dot matrix printer.

The artwork can then be scanned or integrated with text in a DTP software package or can be shot with a camera and reduced onto bromide for pasting onto the final artwork of the document. It is not only quicker than setting up a mock-up of the screens, but is more accurate since it represents precisely the screen contents at any given time. Any processing going on at the time the screen dump is activated is halted and resumes once the output is complete. An example of this screen dump output is shown in Figure 5.5.

You should note that a further advantage of this utility is that it can reproduce reverse video characteristics as well as line graphics, something which the in-built print screen facility of a PC will not do. See Appendix G for details of how to obtain the print screen utility which is produced by Bates Associates of Wigston, Leicester.

SUMMARY

Illustrations are, more often than not, a considerable aid to the text of a technical document. Do not forget to consider using illustrations

ILLUSTRATION

Equivalent Manual Process
- Check Stock Cards for availability
- Check Price List for selling prices/discounts
- Check purchase order files for on-order details.

What Next?
If demonstrating single-location stock, go to section 1.4.

If demonstrating multi-location stock, go to section 1.3.

See Chart.

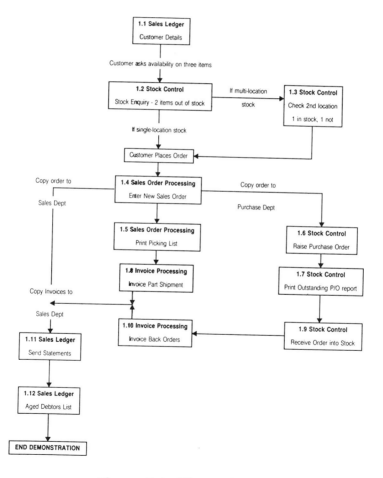

Figure 5.6   Flow Diagram

to represent concepts as well as just the item being documented. For example, a flow diagram showing where the reader of the document is in relation to the rest of the documents and what to read next can be useful – see the example in Figure 5.6.

Illustrations will generally fall into one of the following categories:

Line illustrations
— hand drawn
— machine or graphics device output
— scanned from existing original

Halftones
— screened from black and white photographs
— scanned into a computer for integration in text or direct output

Colour process
— at least a four colour process, scanned

Colour process is very unusual in technical documentation, unless the originators have a considerable print budget or because it is essential to the text due to the nature of the item being documented (eg a manual for a colour printer). Consequently, this book does not go into detail about colour illustrations, but the process by which they are reproduced is briefly explained in Chapter 12 on printers and printing.

Make sure that illustrations are clear and consider the quality of the paper being used before deciding the most suitable halftone screen to use for reproduction of photographs. Some DTP packages are capable of handling halftones through scanned input to quite an acceptable degree of quality, and this is likely to improve in a very short space of time as technology in this field advances.

6 Word Processing

There can be few instances now of those involved in the publication of technical documentation not using the facility of a word processor. By now, the word processor is in widespread use in offices throughout industry and the benefits that it brings to documentation production are well known.

If you are one of the few who still do not use this facility, you should seriously consider purchasing a word processor now. There are many word processors to choose from, and they vary in price as much as they do in features. However, all word processors provide the basic benefits of allowing text to be easily edited and updated, removing the drudgery of retyping manuscripts time and time again until they read and look correct. You do not, therefore, need to spend much on a word processing package to gain these immediate benefits.

Apart from the benefits that a word processor brings to the person responsible for the keying of the text itself, and in many cases this is the author of the technical documentation, there are added advantages when it comes to proofing and producing final artwork for reproduction.

If you do not rely upon a trained keyboard operator to key in the text of documents for you (in my own technical documentation department, a word processor operator is employed to key the main body of the documents, though authors still prefer to key most of the draft text themselves as they go along) then you should get to know the facilities of your word processing system in detail.

Many of the features of the more sophisticated word processors can be useful, for example the ability to structure the document's layout, and facilities for checking spelling etc. But choosing the right word processor for your requirements is a difficult task in the current market where there is more choice than you can adequately give attention to.

WORD PROCESSOR TYPES

Broadly speaking, word processors fall into two main categories; dedicated word processors and word processing software on microcomputers. The dedicated word processors are so called because they are machines that are tailored for the specific use of word processing and perform no other task.

These are not recommended for the first-time user in the technical documentation environment, since you will soon be welcoming the use of such software facilities as graphics or desktop publishing, so to dedicate yourself to an inflexible machine is not a wise thing to do. However, the second category, using word processing software on microcomputers, is by far the better choice, since it not only offers you the facility of being able to use the microcomputer for other tasks, but it enables you to expand the word processor's facilities. You may need to upgrade to better and more sophisticated word processing packages as they become available, because you may not know, first time around, whether any particular word processor best suits your needs. As your needs grow, so may your demands on the word processing software package.

The following chapter discusses desktop publishing, but you will find that many of the better word processing packages have features which are more like desktop publishing software to the point where, before too long, the two types of package will converge. What is available only in the desktop publishing package today, will be the standard requirement for word processors tomorrow.

WYSIWYG

Many of the better word processing packages available today incorporate some facility for controlling layout. Some offer the facility, for example, for multiple column text layouts and the integration of graphics from external packages. Word processors that allow true

WYSIWYG (what you see is what you get) screen layouts are closer to desktop publishing packages in the facilities they offer. These packages also offer the added advantage of differing typestyles, including bold and italics for highlighting the text for headings etc, though this also depends upon the output device used.

Some word processors include graphics facilities of a kind. These will vary from the ability to integrate a graphics image from an external graphics package to facilities for drawing lines and boxes to enhance the layout, or for the production of simple diagrams, flowcharts, graphs, etc. The more advanced word processors will, with the appropriate resolution of computer screen, enable you to view graphics with the text on the screen. However, the usefulness of these facilities will depend upon the quality of the output device, so let us take a brief look at printers.

PRINTERS

There are three kinds of printers that are likely to be used in providing the quality required for technical documentation. These are:

— the dot matrix printer;

— the daisywheel printer; and

— the laser printer.

Dot matrix printers, while useful for fast output for proofing draft copies of the text, are also useful for producing final artwork. Many of the more modern dot matrix printers can now offer what is known as 'near letter quality' (NLQ) output, and the fact that they can handle graphics is a bonus too. In fact, the dot matrix printer is capable of simulating different typefaces for enhancing the text and adding highlights such as bold headings, and generally provide more in the way of layout capabilities than the daisywheel printer.

For a better quality of image, ink-jet printers are now becoming quite popular alternatives to the standard dot matrix, and while they rely on loosely the same principle for making up the image, they are quiet, and can give better results, particularly where graphics are concerned. Figure 6.1 shows an example output from a word processor with a dot matrix printer.

The daisywheel printer is restricted by the typeface of the daisywheel itself, and does not have any graphics abilities. Generally speaking, however, the letter quality output of the daisywheel is better than that of the dot matrix printer, and therefore provides better quality camera ready artwork if it is to be used for final artwork production or for photocopying. One important consideration may be cost, and daisywheel printers are generally more expensive than the dot matrix variety, even when compared to ink-jet printers. Apart from that, if you wish to consider your working environment, then daisywheel printers are considerably noisier too. Figure 6.1 shows the difference between the dot matrix output and the daisywheel, giving you a comparison of how each type can reproduce in print.

Laser printers are now driven by many of the popular word processors. Used in a technical publications environment, they can turn your word processor into a device for producing very suitable artwork for reproduction. Even without the sophisticated facilities of a desktop publishing software package, the word processor and a laser printer can be adequate for many technical documentation requirements. Figure 6.2 shows a sample from a technical handbook

4.2 Write Protect Notch

Each diskette is provided with a `write protect' device which can be identified as the small square cut out of the right-hand side of the diskette. When this is covered with one of the self-adhesive tabs provided with your diskettes, it will not be possible to record information on the diskette. This would be particularly important, for example, when copying diskettes. If the wrong diskette is placed in the drive to which you are copying the information on the diskette would be destroyed if the write protective notch is not covered. Covering this notch will avoid such unfortunate occurrences. Care must be taken not to cover the notch on diskettes which are currently active however, as this will prevent the system from recording information.

4.2 Write Protect Notch

Each diskette is provided with a `write protect' device which can be identified as the small square cut out of the right-hand side of the diskette. When this is covered with one of the self-adhesive tabs provided with your diskettes, it will not be possible to record information on the diskette. This would be particularly important, for example, when copying diskettes. If the wrong diskette is placed in the drive to which you are copying the information on the diskette would be destroyed if the write protective notch is not covered. Covering this notch will avoid such unfortunate occurrences. Care must be taken not to cover the notch on diskettes which are currently active however, as this will prevent the system from recording information.

Figure 6.1 Printer Outputs

WORD PROCESSING

12 Enquiries

Select function 5 (Enquiries) from the payroll processing menu.

The screen will display a box in which you should enter the employee number to whom you wish to enquire, and press enter.

You will be asked if you require a hard copy (printed copy) of the enquiry. Reply 'Y' or 'N'.

The printout may be directed to the printer, to the screen or to a disk spool file (although on this occasion there is no value whatever in directing the printout to the screen). Press the enter key if you wish to change the destination of the printout. Each time you press the enter key, you will see the selected destination of the printout on the bottom line of the screen.

If the printout is to be directed to the printer, ensure that it is loaded with plain listing paper, that it is aligned to the top of the page, and the printer is switched on-line.

Press the space bar to continue.

The enquiry will be printed on the screen as well as the printer and will show the following details:

a) Employee name and number
b) Analysis code
c) Tax code and N.I. number
d) Payment method
e) Cost centre
f) Brith date

Figure 6.2   Laser Printer Output

that was output from a word processor on a laser printer. Whilst the resolution does not match that of true phototypesetting, it is often regarded as sufficient for technical documentation, as opposed to marketing literature where quality considerations are paramount.

USING WORD PROCESSORS FOR PHOTOTYPESETTING

You can use the word processor as a front end to a phototypesetter. By embedding codes within the text to specify the typesetting parameters, disks from word processors can be read by many interfacing typesetting bureaux for translating into typeset output. For details on this facility, known as 'pre press generic mark up', read Appendix F on interfacing typesetters.

LAYOUT AND WORD PROCESSORS

Most word processors offer features for laying out your text to make it presentable. Apart from the usual features that are available on typewriters (tab settings and margin controls) word processors (and indeed some electronic typewriters) offer additional benefits in the form of automatic indentation, positioning (centred, ranged left or right, or justified), and multiple column options. For technical documentation, justified text generally looks tidier.

Make use of whatever features there are for text highlighting (bold, underline, etc) with reserve. As explained in the chapter on design and layout, it is easy to go overboard with such features simply because they are there to use. Figure 6.3 shows an example of the sensible use of word processing features for laying out a page from a technical document.

WORD PROCESSORS FOR INDEXING

Your word processor may be a useful tool for indexing your documentation. Whilst some word processors have built-in features for generating and compiling indices, and catering for paragraph and cross reference renumbering, the most useful feature for indexing is a sort facility. If your word processor can sort a column of text alphabetically, then you have all you need to generate a good index.

Those word processors that offer automatic index compilation are not as attractive as may at first appear. You will almost certainly have

WORD PROCESSING

Chapter 7 CBM Report Generator Page 15

7.0 CREATING A NEW REPORT

7.1 **Select function 2** (Define Report) from the Report Generator main menu.

The system will display the Report Index screen for which you will create a new report definition. If this is the first report, the form displayed will be blank. If you have created previous reports (for whatever module) these will be detailed.

7.3 Having displayed the current contents of the relevant index the system will place the cursor on the next free line in the `Disk Name` column. If no lines are available the error message **`Index Full Press ESC to exit`** will be displayed. You may **enter a 10 character name**, which will be used as the disk and file title and **press return**. Having entered a name the system moves to the description field headed `Title` where you may enter any information which assists you in the identification of the report being designed. (Note: this information does not get printed on the report.)

Press return and the system will then check for duplicate file names and if found will respond with the prompt `Duplicate File Names are not Accepted` before returning to the `Disk Name` field. Providing the chosen name is acceptable the system enters the current date in column `Date (1)`.

7.3 If the new report is the first one on the index, the system requests you to **Load a New report disk in Drive 1** and **press the space bar**. If, however, it is not the first, you are given the option of selecting any of the existing reports as a basis for the new design. If this option is chosen, then you will be instructed to remove the program disk and load the chosen report disk in its place. Having done so, the system copies the existing definition on to the new report disk in Drive 1 before instructing you to replace the program disk in Drive 0.

7.4 If the system identifies that the disk in Drive 1 is not new a warning will be displayed and the system will require confirmation before proceeding. Please note that the program is arranged so the program disk itself cannot be overwritten. If the program has identified that a disk in drive 1 is not new, it carries out a short header function using the existing disk ID (see your Pegasus Introduction manual 4.6). (Otherwise the full header operation is performed using the last two digits of the `time constant` as an ID to reduce the risk of disks with a duplicate ID.)

Figure 6.3 Word Processing Layout

to mark up the text for all items that are to be included in the index as well as identifying section numbers, paragraph numbers, subsections, etc. This involves many additional keystrokes, and it may not be practical to do this when the text is entered, especially if it is keyed in by someone other than the author. This means making a separate exercise of marking up the key words, items, phrases, etc that are to be included in the index. You may find it just as easy to compile the index separately, making a note or keying in from a typescript output all the key words for the index, and making a note of the page number or section number, depending upon your preference, and entering this into the word processor as a separate file or chapter.

Once you have typed the items into the word processor, using its sort facility will save the drudgery of working out their alphabetical order. Once sorted, you can easily identify any entries which may have been inadvertently repeated, and edit the text accordingly.

There is another reason for compiling indices manually in this way. An automatic indexing facility in a word processor can only treat an entry in one format. For example, if the text includes information on the operation of a dot matrix printer, and you want the reader to be able to locate this information through an index, you should really consider under what heading the reader may look for this subject. For example, the reader may look under P for Printers, or D for Dot Matrix, and with an automatic indexing facility, you will get the entry only once. If you compile the index yourself, you can include two entries:

'Printers – Dot Matrix .... page number', and
'Dot Matrix Printers ..... page number'.

This is more useful to the reader, since he or she is liable to find the entry first time if the index is compiled in a logical fashion, and offers more than one alternative for discovering the whereabouts of a certain piece of information. When compiling the index, therefore, you can include all the alternatives you want as you come across them in the text, and let your word processor do the sorting for you. If your word processor does not have a sort facility, use a spreadsheet if you have one, and output the sorted index to an ASCII print file, then read this into your word processor for editing, layout and printing with the rest of the document.

USING ELECTRONIC PROOF READERS

A recent innovation in software has seen the introduction of companion packages to word processors. These include spelling checkers, thesauruses and proof readers. The spelling checkers, and in some cases, the thesauruses, have already become an integral part of the facilities of many word processors, but they can be obtained as independent software packages that are generally more comprehensive than the in-built versions within the standard word processing package.

Spelling Checkers

The spelling checkers provide a useful initial check on the typing and spelling of the text of a technical document. Unless specialist dictionaries are used, or existing English dictionaries updated with the technical terms common to your business or subject, standard dictionaries can be cumbersome to use if they point out every unrecognised technical phrase, mnemonic or product name. Provided your specialist terms are incorporated into the electronic dictionary, they serve a very useful purpose when proofing the text first time round, after it is keyed into the word processor. As an untrained keyboard operator myself, I find spelling checkers particularly useful to overcome the poor accuracy of keystrokes, as the transposition of letters is one of the most common mistakes of users of word processors who cannot touch type.

Spelling checkers operate in a variety of ways, from scanning the text files and flagging unknown words with a control character, to the interactive approach of presenting the user with a close alternative to the typed word in an attempt to offer a corrected word. In this latter case, the dictionary searches through the words character by character.

If a word typed as 'techbical' is found by the spelling checker, then it may find the nearest match as 'technical' which you can accept as a replacement within the text. If the nearest match is not the word you are looking for, then you may have to type the replacement yourself. Words that are not in the dictionary, but which are valid because they may be technical terms common to your type of documentation, may be either appended as a new word to an existing or specialised dictionary, or simply ignored, if they are infrequently occurring words.

The problem with spelling checkers which flag the unknown words with some special symbol or character code, is that they require a second scanning by the user to find the marked word and then change it. If the flagged word is in fact correct, the flag character may have to be removed manually, and this can be tedious.

If your word processor has a spelling checker, then you should use the facility before you print drafts of the text at each stage, to iron out the silly mistakes. Under no circumstances will a spelling checker act as sufficient proofing for spelling on its own, since, apart from the fact that some I have encountered are unreliable, they cannot check that the right word is used in the right context. For example, the phrase 'Set the printer dip switch to an', where the word 'an' should read 'on', will not be noticed by the spelling checker, since 'an' is spelt correctly and will be assumed to be alright from simply a spelling point of view.

Thesauruses

I am not sure that the thesaurus facilities provided by some word processors are of much benefit to the technical documentation author. However, they are offered as built-in devices for assisting the word processor operator in finding alternative words when they get stuck for ideas. Whether they are more efficient than having a printed thesaurus to hand is a matter of personal opinion, and if you think it will help you structure the documentation better, then you may prefer a word processor that includes such a facility. Remember that both dictionary files and thesaurus data files use up disk space. The more comprehensive they are, the more disk space they use.

Proof Readers

A more recent development in software to assist authors is the 'expert system' proof reader. These packages claim to be able to check the text of a document to ensure that both the spelling and grammar are correct. The effectiveness with which the document projects its message is also measured by the software, and marked-up output is generated for you to action. Such proof readers are developed

WORD PROCESSING

mainly as an application aid for business and technical writing. The producers of such software claim that by the use of proof reading aids, the author can improve the content of documents to ensure that the way the information is conveyed is strong and clear to the reader.

However, these artificial intelligence analysers have a long way to go before they become really useful as proof readers, and many of the 'mistakes' they bring to your attention are somewhat subjective in any case. However, they do double as spelling checkers too, and if you don't have one on your word processor they can be of benefit, as well as being fun to use until the novelty wears off.

The one I bought to try out creates a copy of the document being processed, with inserted comments pointing out ways to improve the writing. This marked up copy points out possible errors of grammar, style usage and punctuation.

The marked up copy also contains a summary which gives an overall critique of the document; this contains measures of the reading level grade (sometimes referred to as the fog index), the strength of delivery, and the use of jargon. Comments are included about the document's readability, sentence structure and tone. At the end of the summary a list of words to review are provided and these include all words unknown to the spelling checker. Unfortunately, you have to scour through the document yourself to find out where they occurred if you need to put them right, and so it is not particularly helpful in this respect. However, the list also includes frequently misused words, uncommon words, negative words, slang and jargon.

The method used for such a proof reader is that the document is created on a standard word processor, and once saved to disk, the text file can then be scanned by the proof reader and an output file is created for the marked up version of the original. This means that the original file remains intact, and you have the option of amending either the original, based on the comments in the marked up version (in which case you may wish to check the amended original a second time) or you can amend the marked up version, and get the proof reader to strip out all the mark ups to produce a second version of unmarked text, leaving the original intact.

As far as style goes, the proof reader is particularly hot on the use or the misuse of the 'passive voice'. If you are not familiar with what that is, the example in the manual that accompanies the proof reader points out that in passive voice, the subject of the verb receives the action. For example:

'The book was read by Susan'
'The screw should be tightened'
'The ball was played with by the boy'

In the passive voice, a past participle (eg read, tightened, played) follows a form of the verb 'to be' (eg has been, is, are, was). This is apparently considered a weak form of writing and the suggestion is that business and technical writing should be written in the active voice, and the following alternatives to the above examples illustrate the difference:

'Susan read the book'
'Tighten the screw'
'The boy played with the ball'

If you consider such a pedantic approach to proof reading to be important, then you will find such software an invaluable tool. If you are not used to writing in the active voice, then you may quickly get fed up with the fact that your documents are marked down heavily by the proof reader, when it comes to judging their strength, because of your frequent use of passive voice. The repeated comments to this effect become a bit of a bore. Some of the features are more useful, eg checking for repeated words, incorrect punctuation, weak and wordy sentences, etc; but my proof reader did fall down in many cases where I would have expected it to point out mistakes, for example, it did not recognise the errors in the following sentences:

'The engineer are working on the new model'
'The engineers is working on the new model'

Overall, these electronic proof reading tools can be of some benefit, provided that they are not used as a replacement for human proof reading. It is fair to point out that the documentation for the proof reading software included a note which read:

'The program never presumes to make decisions, only recommendations. The final decision is always yours'

Probably one of the most useful aspects of the proof reader I used was that it forced me to examine the structure of my text more closely

```
A.MARLOW/ELITE
DATE 27th April, 1988
PAGE 40
        6.2 Selecting From Menus

        Two  methods   of  selecting from   menus  are  provided
        within Elite.    By default, the   first item on  any
        menu   is highlighted   with a   reverse   video band.
        This "menu bar" can be moved to select other items
        with  the cursor-up or  cursor-down keys.

        By   experimenting   with   these   cursor   keys,   you
        should   notice   that, when   you   reach either   the
        first or last item on a menu, a further depression
        of the cursor-up  or down key will  scroll the menu
        bar   to   the   last   or   first   item   of   the   menu
        respectively.

        Once the   selected item is   highlighted, press the
        return or  enter key to  load that part of the Elite
        system.

        The   alternative method of   menu selection   is for
        the more experienced   Elite user.   A two character
        code   mnemonic can   be   entered at   the   menus, to
        specify which module   of the Elite system is to be
        loaded next.    For example, typing SD and pressing
        return   will  call   the   "Select Data   File" routine
        without the need for selecting options from menus.
```

Figure 6.4   Example Draft Printout

than perhaps I otherwise would have done had I simply read through for spelling. However, the author is never the best person to proof read his or her own material anyway.

DRAFTING COPIES ON THE WP

As a final point on the subject, a word of advice on printing draft documentation from word processors. When printing a working copy of your technical documentation, use the formatting facilities of your word processor to produce the draft with the following characteristics:

a) Set the text width (using margin controls) to no more than 60 characters. On standard 80 column paper, this will give you a wide margin either side of the text (as long as you align the paper in the printer correctly). These margins can be used for making notes and corrections to the text.

b) Print the document in double line spacing to allow for clear reading and insertion of correction marks.

c) Use automatic page numbering and dating facilities in conjunction with running headers or footers, if available, to include the following information on every page of the typescript:

>Author Name
>Title of Document
>Date
>Page Number

The illustration in Figure 6.4 shows an example draft printout for a document from a word processor using the above suggested layout. You should find this a useful format for proofing, and try to adopt it as a standard for all documentation jobs. The date is useful to identify the latest level of documentation if you don't already use some document referencing or levelling system, and the page number is important in case you drop the typescript and all the pages get mixed up!

7 Desktop Publishing

INTRODUCTION

Desktop publishing has already established itself as a formidable newcomer to assist in the publishing world. Employed to carry out every task, from producing a simple form for internal use in a company through to the setting of complex documentation which incorporates both graphics and text together, there is no doubt about the usefulness of a DTP package. Indeed, the technology contained within this software and hardware combination, and the rate of advances being made with the addition of facilities means that DTP is most definitely here to stay.

The decision as to whether to use financial resources to buy in a DTP system is nonetheless a daunting one to make. Like so many consumer items today, there is more to choose from than our brains can comprehend.

The terminology used in this field is particularly confusing. Many of the terms are misused. For example, you will find conflict of opinions between the use of the term Desktop Publishing and other terms such as Corporate Electronic Publishing, Electronic Pre-press Publishing, Computer-Aided Text Processing and so on. Each of these terms could refer to any one element of a market in which the aim of the process is to provide a means of outputting text in a high quality finished artwork format. You will encounter those suppliers who believe their products to be upmarket, that is to say aimed at the serious publisher, and they will describe their products as Electronic

Publishing Systems, in order to suggest a higher level of function from the more popular PC based desktop publishing systems.

However, it is unfair to categorise the products by suggesting that any one solution is best suited for a particular type of work or job. There is a good deal of overlap between all the products, but their differences lie in the direction in which they have moved from their original place within the market. The early stages of the evolution of the desktop publishing market was very much dominated by the Apple Macintosh system and its Laserwriter print output device (see Figure 7.1).

This could run such DTP software as Page Maker and MacAuthor — even MacWrite and MacDraw could be considered desktop publishing software tools. Then packages such as Ready Set Go and Superpage, among others, paved the way for new developments and stiff competition. Apple's market lead was very much as a result of the successful use of high resolution graphics displays which enabled the screen to display the style and format of the output very clearly, before the results were sent to any peripheral device. Also, Apple could boast a very 'user-friendly' interface which seemed ideally suited to this type of application.

After the development of the Apple system, the style of interface was copied by other PC manufacturers and software houses in order to emulate the same high resolution display. Thus the market has been quickly flooded with a very wide variety of desktop publishing systems, all with much the same function and more or less the same flexibility.

Their variety in features often match their variety in price. At the other end of the spectrum, the typesetting houses and manufacturers of phototypesetting equipment began to 'download' their photo-composition software into the PC environment. The fact that composition software, produced for mainframes or dedicated typesetting front-ends, tends to have a higher level of functions for controlling typographical features, means that there is a wide gap in facilities when compared to PC developed desktop packages. This has already begun to be filled as desktop publishing systems are enhanced and developed.

Whilst the PC based desktop publishing systems are learning the

DESKTOP PUBLISHING

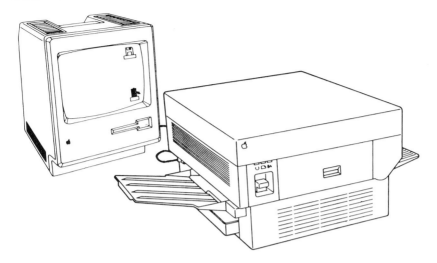

Figure 7.1 Apple Macintosh and Laser Printer

tricks of the trade from their 'professional' competitors, the major suppliers of electronic publishing systems are incorporating such features as high resolution displays, preview facilities and interactive text and graphics page make-up, hitherto the domain of the PC environment only. The Macintosh system has had a long standing major advantage in its ability to have a high level of integration with all the application software packages it runs. Through its universal 'clipboard' and 'scrapbook' features, text and graphics can easily be moved between documents, providing an electronic 'cut and paste' facility which saves on paper, glue and especially time.

As well as working towards their own desktop publishing solutions, the established suppliers of reprographic equipment, such as Compugraphic, Linotype and Monotype, have sought to integrate the Apple system into their own publishing systems.

WHO ARE DTP USERS?

PC based desktop publishing systems are primarily aimed at the business users in companies who have some publishing requirement. Consequently, there are very few companies who can be left out. Most produce some form of printed output: bulletins, newsletters, in-house magazines, staff manuals, maintenance contracts, sales material, training course notes, price lists, catalogues, brochures, advertisements and so on, and all these are potential fodder for a desktop publishing system. For these requirements the desktop publishing system is an attractive solution but, used by the novice, enables the company to produce very high resolution, but terribly badly laid out output.

There is already concern within the market over the pros and cons of desktop publishing and whether the availability of publishing systems has simply dissolved the skills of the graphic artists, typographers, designers and printers.

DTP systems will naturally be attractive to any publishing or authoring department — especially technical publications, where text and graphics often need to be integrated and lengthy documents of varied format have to be produced. If you work in an organisation which already employs the use of desktop publishing for other purposes, then the technical documentation department should certainly look at the system and its many facilities.

For example, the DTP system may provide a very flexible illustration tool which can certainly save time and money in hand drawn illustration work. For small technical leaflets, technical notes and updates it can present high quality, easily readable output which may be of greater perceived value in the hands of the recipient than the commonly word processed and photocopied technical notes. However, when it comes to the production of complete manuals or complex documents, the desktop publishing system may not be so suitable.

Many desktop publishing systems offer what are called style sheets or document formatting facilities. These enable a page design to be set up which can be carried throughout the entire document, and have text imported with spaces or frames reserved for graphics and picture areas. However, some systems suggest that to make best use of these style features, the text of the document should be keyed in through a conventional word processor, and then transferred where required into the appropriate place within the DTP system. This results in a distinction between page make-up software and a DTP system which effectively handles both make-up and word processing. This does not mean that page make-up software cannot allow interactive text entry, but that it is more convenient for larger documents to be typed into a standard word processor. You should, therefore, take care when choosing your system to check that your word processor is compatible with the page make-up software.

On the other hand, in a system which allows you to key in all the text, and format its layout as you go (as indeed many page make-up packages will allow) the processing time may be noticeably slow. Early versions of some DTP software packages were laboriously slow in scrolling text on screen because of the amount of display formatting that had to be processed. This made keying in of long documents tedious, especially when drafting. If the software had no facilities to import text from a conventional word processor, then the user was faced with proofing every draft in near typeset quality.

IDENTIFYING YOUR NEEDS

You would do well to identify straight away a DTP system that suits your needs. If you do intend to purchase a system, especially if you are having to budget for one or justify the expenditure, be prepared to find that soon after the purchase of the initial system, you find that you need to change. If you are lucky, this may only mean upgrading the system's software to incorporate some new features that have since been added to the system and which were lacking in the system when you bought it. If you are not so lucky, you may have to completely scrap the system you have, and start again. This is particulary expensive where peripherals are concerned, especially laser printers.

The best advice is to choose the DTP driving software first. If this comes close to doing the job you want at an early stage, then you will have done well. Check what hardware it will operate on and find out if there is a version which will operate on any hardware you may already have. For example, if your company uses PCs, and your department just happens to have an IBM PC for word processing, you would be advised to seek a DTP software package that utilises this hardware. You may need to make some modifications though. The screen may not have a high enough resolution or the memory in the machine may not be enough to run the DTP package. These DTP packages are very hungry and eat up lots of memory — this is because so much processing is required for the complex screen handling necessary to display the true 'what you see is what you get' displays, plus any graphics that you happen to throw in.

So additional costs may be incurred to upgrade existing hardware to suit the needs of the DTP software. Alternatively, you may take a liking to a specialised piece of hardware that you would dedicate to the task — but beware! If your company employs the use of PCs or DOS based micros and you bring in an Apple Macintosh, you immediately limit your flexibility when it comes to change.

If, after some use, you find that the DTP package has a shortcoming, and you have since discovered a marvellous new DTP software package that has the very features you are looking for, don't be surprised if they are not compatible with the hardware you've got. If not, you have a very expensive exercise ahead of you to rectify the situation. While if the reverse case were true — you had bought a PC because the company used PCs throughout and you found you had to change to Apple systems to do what you want, it may be easier to find a home for the PC in some other function which would at least integrate with existing systems without any mismatch of protocols.

MAKING GOOD USE OF DTP

Having seen the seductive qualities of desktop publishing software demonstrated to you by computer dealer or at an exhibition, you will no doubt be eager to get your hands on the keyboard and try out the many features. So, before we discuss the disadvantages, here are some of the good qualities which will definitely enable your publications jobs to be more effectively produced, provided you make proper use of the system.

For one thing, you have complete control over the layout of the publication. You can experiment with differing designs and layouts, typestyles and graphics, and see the effects immediately displayed on screen or output to some high resolution peripheral like a laser printer. This is not only fun to do, but has considerable cost advantages over sending marked-up copy typescripts to conventional typesetting houses and waiting for the results. This puts the responsibilities of the designer firmly in the hands of the DTP operator, who may also be the author of the technical publication, and this is a very attractive enhancement to one's job characteristics.

DESKTOP PUBLISHING

The ability to output the final artwork, or even print the finished product if you use a page printer as the printing device itself, provides the additional benefits of speed of throughput and cost savings against studio work and printing.

By using a laser printer, the desktop publishing system offers you the facility of producing near typeset quality output in your own working environment for something less than £5,000. This is high quality at a relatively low price. The laser printer can emulate real typefaces as well as daisywheel printers, is silent (or nearly so) and quite fast. Many laser printers (those known as page printers) can output at the rate of between 8 to 12 pages a minute.

One of the attractive aspects of desktop publishing is certainly the self reliance that it offers publications departments. So much more can be done by the department in terms of controlling the publication of technical documentation, than could be achieved through any other means. Traditionally, this would involve additional resources and capital costs in-house. It seems a natural progression to take the word processed text of a technical manual, and add the format and style required for the finished output of the same document, and then print the artwork directly. If used in conjunction with graphics software, scanners and the like, the possibilities seem almost endless, as text and graphics can be integrated together, layout checked on screen first, and then printed in one single process.

Although laser printers can offer the printing facility as well as the artwork production, this will normally only be practical for small print runs, especially where desktop laser printers are concerned. In most cases, work output from a desktop publishing system still needs to be printed through conventional processes. This is even more necessary if the final document is to require colour processes and large print runs.

In general, it has been the advent of the cheap desktop laser printer which has enabled desktop publishing to be so successfully accepted into business and industry as an in-house system. Although many DTP systems provide means of driving phototypesetters, the majority of installations are based on the use of laser printer output. This has seen a gradual change in trend, however, as cheaper and more compact phototypesetting units have become available, like the Monotype 512 and the Compugraphic 8000.

Nonetheless, phototypesetting is still regarded as expensive and laser printers at the cheapest end of the market are now costing no more than good quality daisywheel or ink-jet printers. If you are not familiar with laser printers, below is a description of how they work in principle.

HOW A LASER PRINTER WORKS

The basis of the laser printer machine is not far removed from a conventional plain paper photocopier. The layout of what is to be printed is controlled by the 'page description language' (see Glossary, Appendix I). This builds up an image of the page to be printed using what is known as a raster image processor, in a technique called 'bit mapping' (see also Chapter 5 regarding scanners). The laser printer uses the 'bit image' to transfer, through a laser light, the page onto a light sensitive drum. The drum itself holds a positive electrical charge when it is rotated, and the laser draws the picture on the drum as it scans, by means of neutralising the charge of any part of the drum where a 'bit', or black dot should be. This leaves the remaining surrounding areas positively charged. Where there is positive charge, will be the white areas of the page (see Figure 7.2).

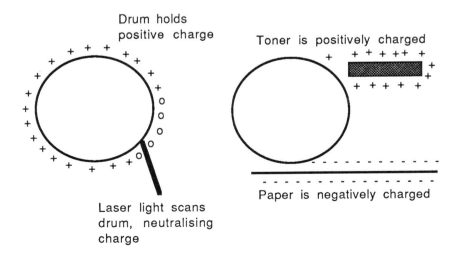

Figure 7.2 How a Laser Printer Works

A toner cartridge in the printer contains the black toner used to reproduce the image, and the toner itself is also positively charged. The drum of the laser printer revolves, picking up the toner in only those places where the black is to be. This is because charged particles of the same charge repel one another. Hence the positively charged toner is repelled from the positively charged areas of the drum, but will stick to the zero charged areas where the bits or dots of black are to be reproduced.

As paper is fed through the laser printer, the printer gives the paper a negative charge. Since the charge is opposite to the positive charge of the toner on the drum, a simple law of electrostatics dictates that the toner will be attracted to the paper. In the final stage, the paper is passed through heated rollers which fuse the toner particles to the paper at a very high temperature. The bit image having been transferred and sealed onto the paper, the output tray of your laser printer presents you with a faithful reproduction of the image originally bit mapped. The higher the resolution of the laser printer, the more precise will be the quality of the output. The resolution is measured in the number of dots per inch that can be processed. In general, most current laser printer technology offers 300dpi, which, although good, still does not match the 1200dpi or more of even the low end phototypesetters. There are already some 600dpi laser printers finding their way into the market at suitably higher prices, but these will shortly become more widespread and the cost will drop correspondingly as technology improves and production increases. Nevertheless, the method by which laser printers operate make them a very attractive alternative to phototypesetting which uses a different process, and is explained in the following chapter.

CAUTIONS FOR DTP USERS

While it can be said that desktop publishing offers a great deal to the technical publications department in terms of freedom of control over design, and the ability to cut production costs, there are disadvantages to consider too. It should be clarified, however, that the disadvantages do not necessarily apply to all, for they depend upon the background and experience of those using the desktop publishing systems.

Given a trowel, bricks and mortar will not enable me to build a house, or even a reasonably good wall. Such tools are only at their most effective in the hands of a skilled bricklayer. The computer software market, being a highly competitive industry, spends a great deal of time, effort and money on the marketing of the products which saturate every gap in the market. Desktop publishing is no exception to this. Indeed, it has been so dominant in the last couple of years, that it has surpassed all previous attempts at technological 'hype'.

Some vendors of desktop publishing software have misled users by suggesting that the possessing of a desktop publishing system is going to improve their corporate appearance on paper, and that it will solve their documentation production problems in-house, in one go. Furthermore, it is suggested that anyone who can use a word processor can use a DTP system, since it is regarded as 'a natural progression from word processing'.

You would find it absurd if a vendor of building supplies suggested that by the supplying of tools and materials necessary to construct a dwelling, in a cost effective package which included a manual, that you could not only build your own house in the cheapest way, but it would look beautiful too!

Given a software package that possesses the elements of design, provides the tools for typography and the facilities of graphics, it is not the layman who can produce beautifully presented documents, manuals, reports, etc. Desktop publishing software has great potential, but must be used with care so that it is not abused technology. The traditional skills of editors, graphic artists, designers and typographers are still important and relevant, even in the face of this new 'easy to use' technology.

Some packaged desktop publishing software now provides training in the form of guides to design and layout, fundamental information on typography and even basic design templates to help the user produce reasonably good quality results.

Should you decide to incorporate a desktop publishing system into your technical publications department, then you must consider what experience and knowhow you already have to draw on, to produce good results from such a specialist tool. If you are used to word processing, the desktop publishing system may appear to be an extension to that process. But this is only partly true. The difference between good and bad output from DTP has much to do with the user's understanding of what the system is best suited for.

You cannot expect to produce professional results immediately. If you have little design experience, sometimes the accompanying documentation can help in this direction. There are courses on desktop publishing and design too, that can point you in the right direction. Books on the subject are now available, and these will help on matters of selection of the right system, and provide you with some background knowledge on design. Only when the users of DTP are educated about what is involved will this new area of technology produce really effective results.

In order to overcome the potential difference between the skills required to make the best use of DTP, and those possessed by the user, some DTP software vendors are building 'expert systems' which control the aspects of typography in an effort to impose or instruct the 'rules' of good design and layout for the document being processed. Not surprisingly, development in this sphere is paramount in the USA, and artificial intelligence for design work is well past the formative stages. Generally speaking, however, such efforts are being directed more towards design for magazines, newspapers and advertising, and there is little guidance of this sort within the reaches of technical documentation authors or publications managers, that is of relevance to technical publishing.

If you are a manager of a technical publications unit, one of your considerations must be about who gets to design the layout of the pages. If your authors are keying in the text of the document as it is originated, how much control do you think they should have over the layout? Authors themselves may be delighted at the prospect, however, and there may be an advantage in enhancing the job specifications of authors by allowing them to exercise more control over the finished product than they have previously had. In short, installation of DTP brings with it many management problems as well as technical ones. Implementing a house style may be a useful step forward if anyone allowed access to the document can affect the layout or style. See also Chapter 3 for further details on house styles.

Remember too, that as far as technical documentation goes, your main consideration should be the successful communication of the information in your document to the reader. It is easy to get carried away with the facilities of DTP, but many of the fanciful aspects that may suit the design of your own adverts, for example, are not necessarily any use on a technical document. Keep designs simple.

The capabilities of a DTP system should be secondary to the content of the technical publication, but design itself should never be ignored. A well presented document will be received well, and if the design is effective, it will help the reader to receive the information and promote the reading of the content in the first instance.

Apart from keeping DTP output simple, you should aim for consistency. If your documents contain varied page designs or layouts, they can look jumbled and confusing. A good, simple design will not be boring over a long document, since the reader should not consciously notice the design element as such. If the reader's attention is immediately distracted to the page layout, chances are that they are not concentrating on the content. If this sounds like a repetition of the points made in the chapters on design and layout, I make no apologies for this because such pitfalls are becoming more often attributable to those users who are busy playing with the features of DTP rather than concentrating on good clear design. This is largely a result of misleading marketing on the part of the vendors of DTP solutions, but you have been warned!

8 Phototypesetting

INTRODUCTION

The facility of phototypesetting has been around for many years. Generally, any work that required phototypesetting had to be handled by a typesetting bureau or printers, since the equipment was of a much too specialised nature to be considered as a viable in-house facility. Copy for phototypesetting has traditionally been rekeyed by the compositor, and still is for many print jobs and book publishing. This makes for a labour intensive and therefore expensive operation; as a consequence, phototypesetting has been reserved for publications that demand high quality presentation, such as sales promotional literature, books and the like.

For the producer of technical documentation, the need to provide a complete and accurate document is generally more important than the need for expensive, high quality presentation. With many technical publications departments producing material which has short print runs, the expense of phototypesetting does not warrant its attention. Even within government organisations, the quality of typed and photocopied technical documentation is readily acceptable, and since cost considerations are often paramount, phototypeset artwork would not be expected unless the publication is for public distribution.

Where very large print runs are involved, the initial cost of typesetting can often be budgeted for on the basis that the overall production costs, when spread over the large number of copies printed, brings the cost per document down to an acceptable level, and the added benefit of high quality presentation is economically feasible.

Much has happened in recent years to change the cost and availability of phototypeset technical documents, to the point where it is far more widespread, and considered important where documentation is used in a competitive situation. By this I mean that technical documentation produced for consumer products is in itself part of the product, and therefore demands attention to quality. The publications that present a high standard of finish will reflect the image of the company producing them. Technical documentation in the computer industry has almost adopted a recognised format of presentation, particularly where software is concerned. This has largely been due to IBM's influence, by providing A5 documentation in ring binders and slip cases containing high quality technical documentation that has been phototypeset. For some time this has set the standard by which other companies in the same market appear to measure their own publications.

With the exception of documentation that is used for in-house data processing departments, engineering workshops, or small print runs of documentation for specialist requirements, almost all product orientated technical publications that do not meet the standards of typeset, printed pages, may be considered by some consumers as inferior.

For software vendors, the appearance of the documentation can play a major role in the success or failure of the sales and marketing, particularly for products that are bought off-the-shelf. While many will agree that 'you cannot judge a book by its cover', consumers are influenced by the appearance of the packaged software product, where the documentation is usually the most noticeable component and often becomes part of the packaging itself. This commercial pressure has forced technical publications departments in the computer industry to seriously consider phototypeset text as an important criteria for success, where traditionally the concerns were concentrated on the content alone.

Added to this, the availability and cost of computer based typesetting has paved the way to bringing this high quality closer to home, and it is readily accepted as part of the production budget for a job. Naturally, desktop publishing, as discussed in the previous chapter, has provided an intermediate stage of quality. Where previously one

PHOTOTYPESETTING 113

had to accept either typed reproduction or phototypeset text, with a wide gap of cost and quality in between, the introduction of laser printing technology has closed that gap. This has had two effects. One is to introduce technical documentation departments to the facility of higher quality presentation, which may have encouraged them towards considering true phototypesetting. This transition may have happened much faster than it would otherwise have done, had a cheaper option not been available between the two ends of the spectrum. The second effect of DTP is to erode the market for phototypesetting. This is because many will happily accept laser typesetting as adequate quality for selling products in the competitive environment, and as technology in the field of laser printing improves, the difference in quality between phototypesetting and laser printing begins to reduce. This will naturally dissuade the advance into phototypesetting from DTP for many technical document publishers.

WHY USE PHOTOTYPESETTING?

There can be no doubt that, whilst DTP originated and laser printed technical publications have made a vast improvement in the presentational qualities of technical publications, phototypesetting provides the optimum quality. Even with laser printers of around 600dpi, this is still at least half the resolution of phototypeset text. Compare the quality of text in this book with the illustration of laser printed output in the previous chapters and you can see the difference. For many technical documents this difference will be unimportant, but in the field of computer software documentation the major players in the market all use phototypesetting. Take a look at any manual for a product such a LOTUS 1-2-3 or dBASE III, or indeed any IBM or Microsoft product, and you have the standards for quality laid out in front of you.

Phototypesetting is no longer as expensive as it used to be. There are many ways in which you can have documents phototypeset without having to go to the expense of purchasing phototypesetting equipment. For a start, you could send your word processor disks, containing the document files, to an interfacing typesetter who will provide you with typeset bromides based on your marked up instructions (see Appendix F on interfacing typesetters).

Alternatively you could purchase or rent a front-end device to a remote phototypesetter, and by means of a communications line via a modem, send jobs down to the typesetting bureau, who will queue and process the work and return the artwork by post. There are a few typesetting bureaux which offer this facility, and you pay for the time that you are 'on-line' and the processing and materials costs. This is considerably cheaper that using a conventional setting house, since you are controlling the typesetter yourself. It is as close as you can get to in-house phototypesetting, since you have full control of the job, but the typesetter is located elsewhere. By means of a proofing device based on a fax machine, you do not have to wait for the job to come back to see the results as the fax machine enables quick feedback to assess the layout of the finished artwork.

If you decide that you can justify the cost of an in-house system, then this will be the most efficient way of producing your documentation artwork in typeset form. The advantage of in-house setting is the speed with which you can see the results of the job, and this can cut down your production timescales significantly, especially if you have to work to tight schedules.

Whichever the option used, phototypesetting remains an important requirement for technical documentation in the computer industry. Although the methods employed to drive phototypesetters are coming more into line with the desktop publishing software, phototypesetting systems generally, as supplied by the major manufacturers, provide a greater degree of control over typographical characteristics, for example, by allowing the fine adjustments to spacing between characters and lines that desktop publishing systems do not yet offer.

HOW PHOTOTYPESETTERS WORK

Figure 8.1 shows a diagram of the principles behind phototypesetting. As the name implies, the process is one which uses light on photosensitive paper to produce the images. The photosensitive paper has to be processed through a chemical system much like film exposed using a camera. The basic units comprise: a front-end system which drives the phototypesetter through the use of composition software, the phototypesetter itself and a processor. The modern phototypesetters use a digitising system to convert the information sent from the front end to control a light source (such as a CRT or laser) which exposes the text images on the paper. The software controls the

PHOTOTYPESETTING

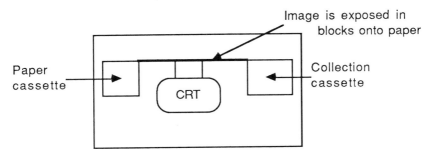

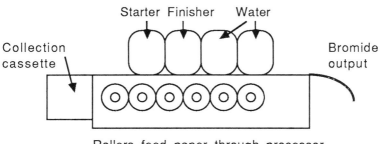

Figure 8.1 Principles of Phototypesetting

typefaces and founts, so the variety of founts available on digital phototypesetters is only restricted by the available software for them. Up to 2,000 typefaces can be available for some systems, though most in-house installations will only have a limited number to satisfy the requirements of the documentation most often processed.

Photosensitive paper is fed into the phototypesetter and is carried by rollers as the light source builds the image in blocks. Once exposed, a carrier or cartridge containing the exposed paper is transferred into the processor, which contains the chemical baths for developing and fixing the image. Some processors have to be used in a dark room, much like film processing, and may require plumbing-in for carrying out waste chemicals and water from the cleaning baths. However, most modern processors are self contained devices that require little maintenance, are compact, use bottles for collecting waste, and can therefore be used effectively in an office environment.

Phototypesetting systems vary widely in price, though modern digital phototypesetters generally offer a similar level of function regardless of the size of the machine, but the speed of processing will vary considerably with price. For in-house use within a technical publications department, there is generally no requirement for fast throughput when compared to a commercial installation, where jobs have to be processed for many clients in short timescales. In addition to the throughput speeds, system prices will be affected by the number of terminals you choose to have attached to the typesetter, and the demand on the typesetter's capacity will often dictate the size of device that needs to be purchased.

9 The Publications Manager

The role of the publications manager is a diverse one and encompasses the responsibilities of author management and print production control. If one was to summarise some of the main duties of the publications manager they may include, among others, the following responsibilities:

a) The recruitment and development of technical writers

b) Initiation and control of documentation schedules

c) Overseeing the design and layout of documentation jobs

d) The budgeting of documentation jobs

e) Print buying

f) Progress chasing

The role of the publications manager may well include many of the responsibilities associated with that of the author-manager. There is nothing particularly unique about a publications manager's tasks compared to any other manager of another area of specialisation. By definition, a manager of any section has to undertake to successfully control the workings of the department, and to develop and use the resources within it to the benefit of the organisation as a whole. For those who are not familiar with what some of those standard tasks are, below is a brief summary.

The publications manager requires no supervision. He or she will be responsible for the functioning of the publications unit all the time (and this applies to the individual author-manager as much as it does to a publications manager with a team of authors reporting), whether present or absent from the unit.

The manager must ensure that all staff in the publications unit are properly trained for the tasks they are expected to perform, and ensure that someone is responsible for every task at every moment. It is the responsibility of the manager to set up a strong team, to ensure that standards and quality are maintained, and to provide staff under his or her control with the appropriate technical training, motivation and sense of responsibility. In addition, work and responsibilities should be delegated, but not abdicated.

The role of the publications manager should also include the following objectives:

— to set aims and goals for the authors and supporting staff;
— to sub-divide and allocate the documentation workload to the appropriate individuals;
— to monitor the progress of the staff and current workload, checking against plans;
— to develop people, or to manage them in a way that enables them to develop themselves.

All of the above could be a list of responsibilities for the manager of any department. They are, by definition, the major tasks of management. Managing people in a publications environment has few unique characteristics which set it apart from the management of a sales department, administration department, and so on.

PLANNING THE DOCUMENTATION

You will need to plan out the content of the documentation in advance of any writing, so that a structure can be referred to throughout the process. Experienced and novice authors alike can have a tendency to try piecing the document together as they go along, putting an order to it at draft stage. This can invite the omission of items of importance if insufficient consideration to the content is given in the initial stages. Also, lack of planning leads to documents which serve only to provide minimum descriptive details about the subject.

For example, a technical manual on a payroll system, which describes what all the entry requirements are for pension contributions, without forethought, may fail to give information about how the system handles pension calculations and makes accumulations for the year-end. While the latter information may not be necessary for the day-to-day running of the system, it may be useful to the user. Only by planning the requirements in advance can such pitfalls be avoided.

This may suggest that those involved in the planning stages may not necessarily include members of the publications department. Planning the structure of a document for external readership requires input where possible from the reading audience. Their input can help you assess the required content of a document before you begin your planning tasks.

There are various ways in which the document's structure can be planned. One way is to first attempt to produce the list of contents for the main chapters. This will help you to put into logical order the main categories of text, then you can break these main sections down into sub-divisions, tackling the content of sub-sections and paragraphs. Planning will also encompass illustration requirements, layout, design and printing. Thorough planning of all factors before the documentation cycle begins will keep the project smooth running. You will need to know how a job is to be presented, for example, before you piece together the structure of the document in detail. There is little point in planning a document that relies heavily on photographic illustrations if neither the budget for the job, nor the specification permits the use of such illustration material. A more experienced author or publications manager will generally need to spend less time planning the documentation, and as house styles are specified and similar documentation jobs are processed, the planning tasks become less time consuming as requirements in this field become well defined.

RECRUITING AND DEVELOPING AUTHORS

The responsibility of any manager who recruits staff for the organisation should not be taken lightly. As a management duty, the recruiting of candidates for vacancies is probably one of the most important tasks to be undertaken.

It is difficult to identify what characteristics or skills make a good technical author. Ask a dozen publications managers to list the attributes for selection, and you will get a dozen different lists, with some common ground but a good deal of variation too. The chapter on the author's role discussed some of these characteristics, but decisions cannot be based on lists of qualities alone. You will have to consider whether the personality of the candidate will fit into an existing team, for example.

Qualifications that are relevant are hard to identify. It may be that an author candidate has attended some formal training or short courses on documentation or technical writing, but you may feel that it is more important for them to be qualified in the subject area about which they will write.

Experience will be of great value in the successful selection of good authors; experience in both the manager making the selection and of the candidates themselves.

Once in the team, the author must be developed to the fullest potential. This may require on-going training, and you should take the trouble to provide a training plan so that the author can improve on current abilities and gain a wider experience of method. An author can often be an isolated individual, getting little chance to meet colleagues in the industry in the same line of work. Seminars and training courses go some way to at least providing that opportunity on a short term, but it is a good idea to get to know how others cope with the job.

The natural progression for authors is into areas of author management or publications management, but with technology changing and developing rapidly, the possibilities for structuring the duties and responsibilities can be more varied. As mentioned in the chapter on desktop publishing, allowing authors to learn to handle basic design principles and then giving them the opportunity to try these out on documents using DTP can be helpful in improving job satisfaction. Authors will be naturally curious about the new publishing technology, and spreading control over technology that you may have in-house provides ways of adding variety and interest to the task of technical writing.

Do not forget to councel your author staff, just as any other

THE PUBLICATIONS MANAGER 121

managers should do to assess the development and interests of the individuals under their control. Not all technical authors want to stay technical authors and one can become complacent about this fact. This may be because the job of technical writing is difficult to get into in the first place, so one may wrongly assume that an author who has struggled to get into the profession is set for life.

EVALUATING DOCUMENTATION

As a publications manager, you should be aware of other documentation within the same field of interest as your own. You should make it your business to know about this, especially if you are providing documentation for products that are sold in a competitive environment.

This involves an evaluation process of both your own documentation as well as that published by others. You should look for both the good and bad qualities in it. Read it for style, tone, consistency and general structure. Ask yourself questions about the level of information within it to see if the approach adopted by someone else has been better than your own. Do not make the mistake of using other documentation as simply an ego boosting trip for your material. Be honest with yourself, and critical of others only where it is deserved. It is easy to make comparisons on the more obvious differences regarding presentation and printing quality, since these are more immediately obvious. It is less easy to read the content constructively for a comparison of the usefulness of the information it contains as compared to your own.

When evaluating your own documentation, check for the accessibility of information contained in it from a reader's point of view. This will involve 'testing' the documentation in a 'user' environment, to see if your level of indexing, cross referencing and section order is satisfactory. These matters can be overlooked when you concentrate on the accuracy of content alone, and even when checking for completeness. A good user document should guide the reader to and around all the appropriate sections of the text to perform its primary task of communicating information.

Finally, once you have assessed the qualities of the content of someone else's documentation, then you should compare the more aesthetic attributes such as page style, typography, use of graphics, colour and materials such as paper, binding and general print quality with your own.

Another way of evaluating your documentation, is to take the time to listen to and gather up opinions from readers. This can be achieved in a number of ways. For example, if you can arrange to have the documentation 'site tested' with the product (if it is a user manual), then the user site can become part of the proofing cycle of the documentation project. The only problem is that a user site testing some software or hardware product may have limited interest in improving the documentation, so without real incentive they are unlikely to provide comprehensive feedback, so long as they can get by with what you have provided.

Another method is to provide a market survey system within the documentation itself for reader feedback. However, for this to work, you will need to plan very carefully the kind of information you want back and the manner in which you form any questionnaires. I have had the experience of instigating a system of incorporating a freepost reader comment card with documentation for feedback. It was quickly realised that the questions on the card were ambiguous, since the responses indicated a variety of opinions regarding their meaning. For example, the manuals included screen illustrations throughout plus illustrations of every printout of the software package that the documentation was intended to accompany. A question regarding how readers rated the use of illustrations caused one reader to indicate that the illustrations provided were obviously helpful, yet other readers commented 'what illustrations?'. It was clear that their concept of illustrating documentation was not the same as ours. As a result, the questionnaires returned had such a variance of opinions that there was little information that could be appropriately analysed to assess whether the documentation was, overall, a successful publication.

An additional factor which affected the success of this survey was that, compared to the vast numbers of documentation units shipped, only a very small percentage of questionnaires were returned. This highlights a drawback of survey systems, that unless the reader has some reason or incentive to complete the form and return it, the task will be regarded as a waste of time on their part.

A further serious mistake I made was cribbed from another market

THE PUBLICATIONS MANAGER

survey form that accompanied a system that I had purchased. This was the inclusion of a section that read 'Please indicate if you have found any errors in the documentation, stating briefly the error and the page number'. I only ever had one form returned that had a response to this question, and it quite rightly stated 'Checking for errors is your job, not ours!'. I guess I asked for that, but it makes you learn to consider the questions carefully before you ask your readers.

A final and obvious form of feedback will be that which is not asked for, but volunteered by the reader or user of the documentation. This is generally in the form of complaint, on the basis that few will inform you that everything is fine. I am glad to say, however, that there are those who do take the time to praise documentation, but criticism is more useful, since it is only on this that you should take action. As a publications manager, if you consider your documentation is the best it could be, then well done, I hope you find another job soon.

PRINT BUYING

This is a task that requires experience. You will need skills that involve you in being able to plan, prepare, budget, order and progress chase print jobs from conception through to completion. If your documentation production involves the use of external sources for artwork production, then the print buying role will include the control of services such as typesetting, illustration, photography and so on. How detailed this responsibility will be undoubtedly depends upon the complexity of the work being produced. If your documentation is word processed, and then reproduced by a photocopying or quick print process in small print runs, then the task of print buying will be particularly limited. Once you get into the realms of offset litho printing, and specialist finishing requirements, then it is part of the role of a publications manager to ensure that, on behalf of your company you are getting the best quality at an economical rate, while, from your readers' point of view, you are offering the best job that your budget can offer.

Print buying is defined as the management of all the resources available to meet an organisation's print needs. To do this effectively, the publications manager must not only be knowledgeable about the processes already employed in production of current documentation, but also other processes and materials available. Doing so ensures that decisions can be made effectively and wisely. Today, print processes are heavily technology based, and it can be quite a task to assimilate the knowledge required to both understand and make best use of the facilities available. You will be involved in buying both services and materials, and there is a great deal to choose from in both categories. You may consider that it is wiser to employ the use of an agent to do the print buying for you, such as a marketing agency with print process knowledge. This has the advantage of easing the burden on your job, and is less time consuming. Your task becomes one of submitting requirements to the agency who then take on the print buying, and possibly even part of the production of the documentation, and then progress chase the job from there on. The disadvantages are lack of control on your part and greater expenses.

The chapter on print processes may help if you are totally unfamiliar with the fundamentals of printing, which you must have knowledge of to do any effective print buying. However, the information supplied is limited and you should look to increase your knowledge through experience. When dealing with printers, take the time to visit their premises, and talk to them about how they do their job. They will not volunteer such background knowledge generally, but are usually quite happy to oblige if you ask. You can pick up quite a lot of information about various aspect of printing by getting to know the printers you deal with and those who you might deal with.

Once you have gained knowledge about what different processes are available to you and what the current technology offers, you then have the fairly awesome task of understanding what types of job suit what processes. You must be able to identify the most appropriate materials and processes for a particular job. Some publications will be suited to production using paper printing plates, others may need processes that use web-offset presses using metal plates and so on. When you are familiar with the differences, you will be able to make better judgements of how best to tackle your print requirements.

It should not be overlooked that many printers will advise you on the best processes anyway. It is their job to know what process will be required for any particular job. However, the publications manager who accepts quotations from printers for various projects

THE PUBLICATIONS MANAGER

simply because they fit the budget, may never get to appreciate what process is being used. In the light of the pressures of the job, it can be easy to leave the task of relating the job to the process to the printer alone, and not to get involved any further. The printer may choose paper, other materials, and a process on the basis of what you intend to pay, rather than what best suits the job. Unless you take time to understand the options for yourself, you will be in no position to change this state of affairs.

Print Specifications

You should have specifications drawn up for jobs that you have printed. The specification should include details of the format used, eg A4 or A5, whether you are providing phototypeset, laser printed or word processed artwork (as this could affect the type of paper), finishing requirements, etc. Figure 9.1 shows an example print specification for a user manual for a software product.

The specification, once defined, can be used to obtain quotations from more than one printer, to help assess the cost of printing and in an effort to reduce costs where possible. The specification becomes as detailed as the complexity of the processes involved, so for a document that is bound in a ring binder and slip case, this should include details of the materials used for packaging, colour, and who is responsible for the origination of artwork.

This latter point is important. Many quotes from printers will, unless you have been specific about the details of the job, only quote the print costs, leaving any artwork processes as a variable additional cost. It is therefore important to identify who is responsible for handling artwork origination. If you are capable of producing complete camera-ready artwork in-house, then you may only rely on the printer for imposition, platemaking, printing and finishing. If artwork has to be handled by the printer, the costs can be affected substantially depending upon their facilities, as the printer may agree to take on artwork production, but contract it out to some external service. It will pay you to know what facilities your printers have in this respect so that you are at least aware of how much they can do for you on their own premises, and how much they will put out to others.

Pegasus Publications Department
Job Processing Schedule
& Print Specification

Job No._____   Job Name_____

File Opened __/__/__   Due Completion Date __/__/__

Details_____
_____

Print Specification_____   Qty Req._____

Type_____   2nd Colour_____

Style_____   Print Finish_____ Paper_____

Illustrations_____   Throw-Outs_____ Binding_____

Special Requirements_____
_____

Status Summary:
	Due Completion Date	Completed	By Whom
Familiarisation	__/__/__	__/__/__	_____
Documentation	__/__/__	__/__/__	_____
W/P	__/__/__	__/__/__	_____
Copy Proofed	__/__/__	__/__/__	_____
Typeset	__/__/__	__/__/__	_____
Illustrations	__/__/__	__/__/__	_____
Artwork to Printer	__/__/__	__/__/__	_____
Page Proofed	__/__/__	__/__/__	_____
Ozalid Proofs	__/__/__	__/__/__	_____
Printed	__/__/__	__/__/__	_____

Remarks_____
_____
_____
_____

Figure 9.1   Sample Print Specification

THE PUBLICATIONS MANAGER

Notice that the print schedule includes details of printing timescales, though you may not be able to specify this, except in individual cases at the time they occur. However, if you know you have very short timescales generally, perhaps because you always get your documentation finished just days or hours before it is required, then you will need to identify this requirement to the printer. Print presses do not have limitless capacities and are usually tightly scheduled to ensure that presses are running continuously for the sake of cost effectiveness. If you intend to interrupt that process, either by demanding very quick turnrounds or by continually putting off submission of copy or artwork because your own schedule is changing rapidly, then this may cost you extra. Remember also, that while printers may be able to pull a few strings when you want something quickly, they may do so by the use of overtime, and the job will be correspondingly more expensive.

Supplier Selection

Selecting the supplier for print jobs is not an easy task to define using any standard criteria. Much will depend upon your individual requirements. Locality may be important, for example, though many print jobs for technical publications are adequately handled in more remote sites.

You will probably experience cold call contact from printers' representatives looking to take on your print requirements. If you agree to arrange for them to visit you to discuss your requirements, be prepared for each one to be the best for your needs in their opinion. Having a clearly defined print specification will help to sort them out as far as price goes, since you can get them to quote on similar job specifications for comparison purposes. Beware of those printers who claim to specialise in printing technical publications. There are quite a number of printers who now offer their services as specialists in this field, particularly with the intention of attracting the many potential customers in the computer industry.

While they may be able to offer you an impressive client list that could include your competitors in the market, and claim that they are geared up specifically for technical publishing, I have yet to discover what special printing methods are used for technical documentation over any other form of publication. At the end of the day, paper will be fed through a press, ink applied and the job will be gathered, trimmed and finished according to your requirements. I know of no 'technical manual' press process that is unique to technical publications, so remember that you may well be dealing with a printer who has no more ability to produce the results than a printer who produces books or similar jobs.

Of course, that does not mean that any printer will do. Some printers may prefer to specialise in high print-run, colour magazine work and may therefore be too expensive for a small print run of single-colour A4 technical documents. This is because the set-up time of their machinery may account for the cost effectiveness on large print runs. In these circumstances, the printers may be unsuitable, even if they agree to take on the job in the first place. Similarly, a two-colour job may be achieved by a printer who only has a single-colour press, but who can achieve the result by passing the job through the press twice. Again, the job may not be cost effective.

The lesson is to find the most appropriate printer for the job itself, but not to be taken in by an attempt to suggest that they have a better type of system for technical publications than any other printer. After all, there are as many ways of producing technical documentation as there are print processes.

JUSTIFYING IN-HOUSE PRODUCTION

Generally speaking, the more you can afford to handle in-house (provided you also have the expertise), the more control, both productive and economic, you will have over your printing requirements. In-house artwork production is becoming rapidly more widespread as the use of desktop publishing systems increases. But printing requirements vary widely in the size and type of work involved and may not be solved, practically, in-house. In general, it is the larger organisations such as government departments, or corporate industries that have the budgets and resources available to employ in-house printing.

Good management is required to ensure that quality is maintained, and unless the appropriate skill levels are also in-house, such departments can be responsible for limited variety of output and poor standards. When taking on any amount of in-house production equipment, whether it be a phototypesetter, printing device or graphics system, the role of the publications manager must be to implement cost control, efficiency analysis and quality control measures, as would be applied to external printing.

COSTING AND SCHEDULING

These two responsibilities are most certainly part of the publications manager's role, and indeed may be that of the author or author-manager if he or she is responsible for ultimate production of the documentation. Two separate chapters are included in this book devoted to these topics.

PROGRESS CHASING

All documentation jobs require constant monitoring and progress chasing. If you are managing a team of authors, you may wish to hold regular meetings to assess current workloads and to discuss schedules. In smaller units or departments, it may be sufficient to discuss current jobs with the individual authors and monitor progress on an individual basis. Whichever method you adopt, your responsibility as publications manager is to know the current status of progress for any documentation job.

Progress chasing will extend outside the publications department to printers and other external sources that may be used. The emphasis must be on maintaining a schedule and then monitoring performance against it. Keep a diary for dates of project activities such as when page proofs arrive from printers, so that all job events are logged somewhere. You might consider setting up a database application on a PC to store details of publication jobs and for logging job progress. Database systems can also be helpful for storing print specification details associated with individual jobs, plus quotes and prices and invoicing details.

STOCK CONTROL

Some publications departments may be responsible for holding the stocks of the documentation and issuing them on demand from the despatch or sales departments. This is often the case where publications departments are responsible for collating and binding the documentation because it is subject to rapid change in short timescales. In these circumstances new information may need to be collated into the current release before binding and despatch.

Whether your department forms part of the finishing process or whether you hold stocks of finished documentation produced externally, then the role of the publications manager may include controlling stocks. This is a difficult area to provide any guidance on specifically, since economic stock control depends heavily on rates of usage, print quantities and other variable factors. You should always be aware of the value of stock held and the rate of usage of documentation, even if another department is responsible for the stock control, since such information may be useful to you in your print buying role.

10 Costing

The costing of documentation projects is necessary to enable budget variance control, and to help analyse performance and expenses, reduce costs and improve financial efficiency. It is also necessary for invoicing purposes should your documentation service be saleable.

To cost a documentation job, you need to break down the project into the components or cost types to an appropriate level. For example, you may consider that, for your purposes, you need only split costs into four categories:

Cost to Write + Cost To Review + Cost to Produce + Cost to Print

Within each of the above categories, you need to identify the associated cost types. 'Cost to Write' will certainly include the cost of the author's writing time, but may also include the cost to word process, the cost of the author's time in learning the subject matter, etc.

'Cost to Review' will identify the costs of proofing and amending at all stages of draft and final copy production, but may exclude cost of printing proof reviews which could be included in one of the other two categories.

'Cost to Produce' could include all artwork preparation costs, illustration costs, etc, and all categories should include their respective expenses and material cost requirements.

There will be a good deal of personal preference to costing methods and there are few basic principles that deserve any amount of detailed explanation, but the following paragraphs provide some simple guidelines.

BUDGETS

Where possible, you should try to allocate budgets for both individual jobs and the cost types. So, for example, a list of job budgets may look like this:

Job No.	Description	Budget £
J001	System Specifications	6,500
J002	User Handbook	8,000
J003	Installations Guide	2,000
J004	Technical Fault Guide	1,200
etc	etc	etc
Total		X,000

The budgets for the cost types is specified in a similar manner, eg:

Cost Code	Cost Type	Budget £
W01	Author 1	15,000
W02	Author 2	15,000
W03	Senior Author	18,000
etc		
	Sub-total Authors	XX,000
WP1	Word Processor Operator	6,500
PC1	Photocompositor	15,000
etc		
	Sub-total Text Processing	XX,000
M01	Materials – Paper	2,000
M02	Materials – Printer Ribbons	225
M02	Materials – Binding Combs	1,200
etc		
	Sub-total Materials Costs	XX,000
OH1	Overheads – Equip. Service	2,500
etc		

In the above examples, the job budgets are allocated to each job as a specification is drawn up in the early stages. The cost type budgets are allocated on an annual basis and then costs are booked to the individual jobs as they arise. The annual budgets can be broken down into period budgets, such as monthly, and monthly variance reports can then be produced. In the case of time costs, these will be booked against the job on an hourly or daily rate basis, then totalled for budget comparison. The materials costs are booked against the jobs to which they apply, but overheads are not associated with the individual jobs, applying to the department costs as a whole. The mix of material and overhead budgets and costs will vary depending upon the detail of cost control you want associated with any given job. In the above example, you may consider costs such as paper and printer ribbons as overhead costs since you do not need or want to break down these costs and allocate them to individual jobs.

Figure 10.1 shows an example budget analysis report for a series of documentation jobs. The variance information is useful to monitor expenditure, and to enable future budgeting to be based on real costs with more accuracy.

COSTING RESOURCE TIME

When costing resource time, you should try to include all costs relevant to that resource. If you want independent cost analysis as a department, you may need to include overhead costs that relate to your department, and those which are associated with the individuals employed. For example, an author's salary may be £12,000 per annum, but national insurance contributions payed by the company should be added to this, since this is part of the expense of employing the author. If the author receives other valued benefits, these must be included also. These benefits may include pension, car, expenses, etc.

If a resource uses expensive equipment on a permanent basis, you may include the cost of depreciation of the equipment used as part of the total resource cost. In these cases, the cost analysis is geared towards costing of the whole department, and the job costs that result should take on the appropriate weighting of all department overheads. You will certainly need to cost jobs on this basis if you intend

```
10.08.88                    Job Cost Variance Report                        Page 1

Job Numbers  Description              Budget    Total    Variance   % Var    %
                                                Cost                         Complete

UM-0031      Printer Manual/Tech. Spec.  2000   1946.25     53.75    2.69    0.97
UM-0042      User Manual Alpha Product   4000   4101.00   -101.00   -2.53    1.03
UM-0043      User Manual Centurion       4000   1009.10   2990.90   74.77    0.25
MB-1014      Product Brochure           12000  10989.56   1010.44    8.42    0.92
PL-0030      November Price List         1500   1585.00    -85.00   -5.67    1.06
PC-801       Parts Catalogue - Issue 4.0 5500   6010.00   -510.00   -9.27    1.09
TS-0113      Technical Specifications     500    395.50    104.50   20.90    0.79
                                        -------------------------------------------
Totals                                  29500  26036.41   3463.59   11.74    0.88
                                        -------------------------------------------
```

Figure 10.1   Example Budget Analysis Report

```
13.06.88                    Time Sheets                                    Page 1

Employee 4005                        Time sheet for w/e 5th June

Job       Cost
Number    Code     Units   Rate    Value

UM-1021   A001     6.00    10.50    63.00   1st draft authoring      5/6
UM-1021   R001     2.00     7.80    15.60   Review draft             6/6
UM-1043   A001     4.00    10.50    42.00   Writing updates          6/6
TS-1132   I012     3.50    10.50    36.75   Illustrations            7/6
UM-1021   A001     5.75    10.50    60.38   2nd draft auth.          8/6
                   -----           ------
                   21.25           217.73
                   -----           ------
```

Figure 10.2   Time Sheet Printout

# COSTING

to charge out your documentation services to a client. You will then need to uplift the costs, probably by a percentage, to make a profit on top. However, for internal costing purposes, all job costs should remain 'at cost' only. In the above example, therefore, a formula for working out the rate cost for an author resource, may look something like this:

(Author Salary + N.I. + Benefits + Expenses + Equipment Depreciation) ÷ (period of charging)

Notice that all items in the formula would be annual costs, which, when divided by the appropriate charging period, say 365 days, would give a cost per day. However, your resource is unlikely to work 365 days per year unless he or she is very keen. Neither does an author work 24 hours per day. Consequently, the above formula has to be adjusted to compensate. You may decide, for example, that an author spends only a certain percentage of time actually writing documentation. It would be unfair, therefore, to charge the full daily rate on a job if the author was involved in other activities.

How you arrive at a suitable cost will depend upon your individual circumstances. You could decide, for example, that approximate costs are sufficient for your purposes, and that you do not want to calculate rates so finely. The important point to remember is to apply whatever rule you choose for costs fairly and evenly to all jobs. Otherwise, you will get an unbalanced view of the costs of the documentation jobs your department processes.

TIME SHEETS

Time sheets should be used if you want accurate cost control of resource time spent on individual jobs. This will be particularly important in a department of many resources which include authors, word processor operators, graphic artists, project managers and the like, all of whom may be working on various tasks at different times. You will need to employ the use of time sheets to gather the data required to place costs against a job, and Figure 10.2 shows an example of a time sheet printout for an author resource.

MATERIAL/OVERHEAD COST SHEETS

Like resource costs, materials which can be identified as being used on specific jobs should be costed to that job. As values are accumulated against the various jobs in progress, you can, at specific periods, compare the cost of materials booked to jobs with the period budget for expenditure on such materials. The same applies to any specific overhead expenses. If you produce artwork in-house, such as phototypesetting or illustration work, you will need to book these costs to the job. If using a DTP system, you may wish to work out a rate cost for the time a job is processed through it, and include the material costs of DTP (eg toner cartridges for the laser printer, paper, etc) as part of that rate.

EXTERNAL COSTS

Apart from the costs of processing jobs internally, you will probably use some external services such as typesetters, quick print shops, printers, and so on. These costs will be easier to associate with individual jobs, since you will probably get quotations first, and the costs are precisely specified. The same applies to purchases of finishing materials such as ring binders and slip cases, which you may have screen printed for your technical documentation. The difficulty is in budgeting accurately for these external costs in advance of the job specifications. Here you will need to rely on past cost information and your experience to make valued judgements.

COST PER PAGE

It can be a useful exercise to calculate the cost per page of your jobs, based on an average of the cost information across similar jobs. You will only be able to do this with any degree of accuracy if your documentation specifications are similar to that of the jobs included in the costing exercise. You cannot compare the cost per page of an A5 typeset and printed manual with a laser printed quick print shop job.

The cost per page is arrived at by dividing the unit cost of the job by the number of pages. The unit cost of documentation can be easily calculated by taking the total job cost and dividing it by the number of finished units. However, this will change with the number of units printed, such that the larger the print run, the lower will be the unit cost. Consequently, if your print runs vary widely, an average cost per page will not relate to individual jobs.

COSTING

The table below shows the following details for each company: Type of Document, (R = Reference manual, U = User manual, B = Brochure, I = Introduction manual, T = Training manual, C = Capability summary), number of pages, writing details (cost, time and whether done in-house), writing cost per page, writing time in pages per week, production/print cost per page, and total cost per page.

Co.	Doc type	No of pages	Writing Cost £	Time m-wks	In H?	Prod./ Print Cost	In H?	Writing: Cost per page	Writing: Pages per week	Prod cost per page	Total cost per page
S1	-	75	6750	5	N	2950	N	90.00	15	39.33	129.33
S3	-	400	-	3	N	16000	N	-	133	40.00	-
S4	-	25	795	0.6	Y	50	Y	31.80	41.67	2.00	33.80
S5	-	50	-	2	Y	-	Y	-	25	-	-
S6	R	600	-	12	Y	-	N	-	50	-	-
S7	U	300	5000	16	Y	15000	N	16.67	18.75	50.00	66.67
S8	U	40	775	7	Y	77	N	19.38	5.71	1.93	21.30
S9	U	120	4700	10.2	N	1900	-	39.16	11.76	15.83	55.00
S11	U	200	2160	4.5	Y	3324	N	10.80	44.44	16.62	27.42
S12	U	100	-	9	Y	-	N	-	11.11	-	-
H1	-	60	-	10	Y	-	N	-	6	-	-
C1	B	4	-	2	Y	-	N	-	2	-	-
C2	-	16	250	0.7	Y	20	N	15.63	22.85	1.25	16.88
C3	I	75	3250	5	N	-	-	43.33	15	-	-
C4	T	200	900	1.6	N	-	-	4.50	125	-	-
C5	U	200	-	5	Y	-	N	-	40	-	-
C7	U	100	3250	2.5	Y	-	Y	32.50	40	-	-
C8	U	150	5750	8	N	5350	N	38.33	18.75	35.67	74.00
C13	C	20	300	2	Y	177	Y	15.00	10	8.85	23.85
C14	U	600	5500	30	Y	1350	N	9.17	20	2.25	11.42
C15	U	48	2500	4	Y	5	Y	52.08	12	0.09	52.18
C16	-	200	30000	-	Y	-	-	150.00	-	-	-
C17	-	150	15000	22	Y	-	N	100.00	6.81	-	-
C18	-	20	-	17	N	-	N	-	1.18	-	-
C20	U	100	13500	18	Y	3000	N	135.00	5.55	30.00	165.00
C22	U	60	3500	8.5	Y	4900	Y	58.33	7.05	81.67	140.00
C23	U	175	7070	8	Y	5755	N	40.40	21.87	32.89	73.29
C24	-	100	11500	10	Y	2600	Y	115.00	10	26.00	141.00
C25	-	75	7000	9	N	6461	N	93.33	8.3	86.15	179.48
No. of respondents for calculating averages:								21	28 *	17	16

* To avoid distorting the figure unnecessarily, four respondents were excluded from the average for the number of pages written per week, so the actual number used for this average was 24. C1 and C18 were extremely low, while S3 and C4 were extremely high. (It seems untypical and improbable that it should take 17 man weeks to write 20 pages, but on the other hand, it seems equally unlikely that 400 pages could be written, reviewed and corrected in 3 man weeks: at 400 words per page and a 35 hour week, this is over 25 words per minute!)

Table 10.3   Page From Digitext Survey

You can test your cost per page calculation on jobs that have already been processed to see if you have worked out a fair cost. If it comes within an acceptable percentage of accuracy of the actual job costs, then you can use this value to budget costs for future jobs, once you have estimated the approximate page count of a document.

Digitext's survey on documentation in the computer industry for 1987, included some information about 'Typical Costs' for documentation. Figure 10.3 shows the result of that survey.

The cost per page varied widely but so did the type of job costed, from a brochure to a training manual. While your own costs will vary widely depending upon the type of publication, and therefore the length of time and effort required in-house, it can be interesting to compare your costs with similar cases, but do not be alarmed if your costs do not match what appear to be similar jobs, unless they are extravagantly out, (notice Digitext's comment at the foot of the table).

11 Scheduling

Planning and scheduling for publications is just as important for technical documentation production as it is for any other production requirement. It is not uncommon to find that the process of scheduling is left to the product development department of an organisation, and that the documentation is simply fitted in somewhere towards the end of the development cycle. This seems to be more apparent in smaller industrial enterprises, but is less so in large corporate or government organisations where there is often stricter control in the areas of both budgeting and scheduling.

If you work in an organisation that doesn't insist on scheduling, then perhaps you should take the initiative to introduce it. For one thing, it should give you better control over the work-in-progress and help you to cope with forecasting requirements. You will find that supplying estimated production times to your colleagues will strengthen their appreciation of your own work and enable it to be fitted in with the general company plan.

BUILDING A SCHEDULE

To build up a scheduling procedure you should start by breaking down into components the various categories of work that form the publication cycle. Don't try to break down functions in too detailed a manner, however, since by trying to schedule events in the cycle in too much detail you will find that the process of scheduling will become too laborious. Indeed, it could become a hindrance to the work throughput itself. Publications cycles which may depend heavily on the development of the product or service can change rapidly and frequently. Therefore, your schedule will need to be a flexible reporting tool. If you keep it simple a certain degree of flexibility becomes in-built.

For example, take a look at Table 11.1 and illustration Figure 11.1 showing a publications cycle broken down into components.

Notice that the items can cover a wide range of jobs but do not necessarily relate to every stage of any individual job. For instance, in the case of writing a manual for some product under development, stages c, d and e may occur several times during the publications cycle, yet occupy different time periods for completion. In fact, if you were working in a small unit or even on your own, items b and c for example, may occur at the same time. The important concept to grasp is that you should find that by breaking down any publications job done in the past into its component processes, you could come up with a different pattern of scheduling of events each time.

Despite the fact that it may contradict any previous methods of scheduling you may have encountered, you are well advised to stick to a single event breakdown which has a global application to all your publications jobs. In any case, when it comes to forecast scheduling you may have no idea of the number of redrafting events that may be required before you can proceed to final artwork stage.

Now look at the example breakdown in more detail.

GENERAL DESIGN CONCEPT

This should cover whatever preparation work has to be undertaken before you put pen to paper (or finger to keyboard these days). This could include an overall design for the documentation, page layout, chapter structure, overview of contents of sections and so on.

It may be as sketchy or as detailed as you like, but if you allow no time for general design concept work before writing the documentation, ie you have the text written then decide how you will present it, structure it, section number it and so on, I guarantee you will not produce the best possible results. At best you will waste time in redrafting and amending. So, if you are planning your documentation projects properly, then you will need to allocate time in your schedule for this. It is easy to overlook, since such work can

SCHEDULING

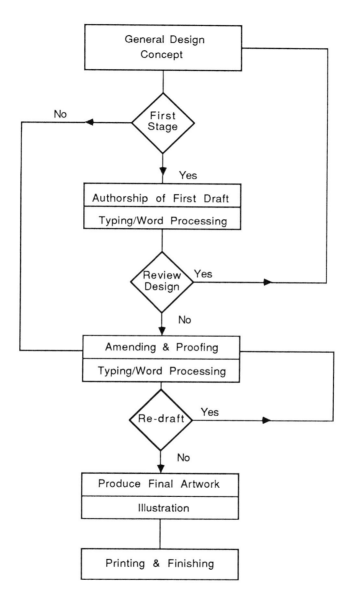

Figure 11.1 Publications Cycle Flow Chart

a)	Design and layout planning
b)	Write draft text
c)	Word process
d)	Review document
e)	Re-draft text
f)	Produce illustrations
g)	Prepare final copy
h)	Arrange print production

Table 11.1 Publications Cycle

be viewed as less productive than actually starting the writing, but the idea that the author can piece it together as he goes along is just not efficient.

AUTHORING FIRST DRAFT

Once an outline of the document has been drawn up the writing can commence and the first draft of the document should be your next stage. The time taken to schedule the authoring of the first draft of any manual, for example, may have to include the learning curve required for the author to understand the subject matter. Consequently, the time period allocated for this would normally be more than any redrafting stage, even if a redrafting stage meant the rewriting of the entire document.

WORD PROCESSING OR TYPING

If this work is undertaken by the author at the same time as the writing, then this should be scheduled together with the drafting. On the other hand, if the word processing or typing is carried out as a separate function then it may also be pertinent to allocate an amount of time for an initial proofing of the typescripts. Indeed this may be a separate proofing exercise to checking for accuracy of content and technical detail.

SCHEDULING

PROOFING AND AMENDING

Proofing cycles may be involved at every stage. If these need resources outside the publications department, for example, one of the development team is used to check technical accuracy, then you should be careful to schedule the time for this. Unless the individuals concerned can be allocated to your schedule at the time they are required, it may be difficult to accurately account for time periods during which the documentation is out of your immediate control. Time allocation for both proofing and amending will have to be fairly flexible anyway, since it is not possible to gauge the amount of effort required prior to this stage of the process.

ILLUSTRATION

It may be possible to schedule illustration work to be done whilst the documentation is still being drafted, provided the job has been adequately planned. If, however, illustration of, say, a software product depends upon the availability of the finished product, then illustrations will almost certainly be one of the last activities prior to publication.

EDITING

This is really a part of the proofing and amending cycle, in that it affects the document after draft stage, but you may wish to schedule it separately if it involves a specific individual's time. For example, if the publications manager has to edit an author's draft before it can be submitted for production, then this additional resource may need to be accounted for. It may certainly affect the cost of the job.

ARTWORK PREPARATION

This is the time allocated for the preparation of the final copy that is used for production, which may be word processed or typeset output. If the documentation has to be contracted out for preparation of final artwork, the timing of the submission and the time allocated for processing may be crucial. Allow for the time that may be required by a phototypesetting bureau, for example, before they can begin to process the job.

PRINTING AND FINISHING

As with artwork preparation, this may be carried out externally to the technical publications department, and will therefore be dependent upon information from the printer/finisher regarding estimated timescales.

After many jobs, the publications manager will become experienced at judging appropriate timescales for most activities based upon historical information of similar jobs and on their own experience.

When scheduling the individual activities, you will need to consider those which have to be preceded by another before they can be carried out. For example, the typing cannot be started until the writing has begun, but it may begin during the time of the writing process; proofing cycles cannot commence until at least first draft, and so on. Once you have established which activities can overlap, you need to associate the activities with the available resources. If you work on your own, then your schedule will be a very simple sequence of events. Where many resources are involved, however,

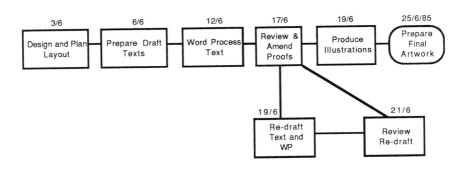

Figure 11.2 Schedule Chart

SCHEDULING

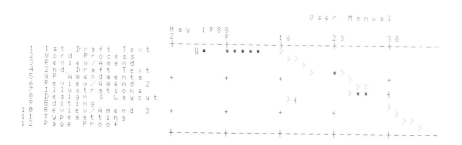

Figure 11.3   Resource Chart Schedule

the scheduling can be a complex task as many activities are being operated simultaneously, and probably for more than one project at a time.

USING ELECTRONIC SCHEDULING SYSTEMS

The use of a computer based scheduling system may be very useful for controlling publications projects. There are many standard packages available which enable project control, resource management and cost schedules. They are all generally similar in the facilities they offer, usually providing you with a means to specify the activities of a job, the resources to be allocated to the activities and the duration. These activities are then built up against a calendar, and having specified a start date for the primary activity, the system can provide you with a schedule of events and estimated completion dates of the various tasks. They are also useful for pointing out where you may have resource shortages or over-allocations.

One of the major advantages of such systems is that they allow for 'what-if' scheduling, and enable the manager or author to amend the schedules easily when circumstances dictate, and see the effect of this on one or a number of current projects.

Figures 11.2 and 11.3 show examples of schedule charts, printed out from electronic project management systems, and show you how useful they can be for providing 'at-a-glance' status reports of documentation jobs. Indeed, publications production is well suited as an application for electronic project control, since it is easy to break down the processes involved. The more control you want over the time schedules, the more detailed become the activities you specify.

Resource tables can be controlled through the project management systems, and these can be useful for indicating the jobs allocated to individual resources and their current capacities for additional workloads. Figure 11.3 shows an example of a resource chart from a system which also provides costing details.

12 Printers and Printing

Understanding some of the principles of printing will help you to both plan your documentation jobs and to effectively carry out the task of print buying. There still appear to be many technical writers who produce documentation copy, which is given to printers to finish, and they do not really know what happens to that copy in the process of printing. This may be because it is not actually necessary to know, but if the trouble is taken to learn something about the process, it broadens the knowledge and provides the background necessary to make better decisions about the presentation of documents.

Printing technology has come a long way since the times of block printing used in China in the 8th century. As far as the Europeans are concerned, it was Johann Gutenberg (c 1397-1468), a German printer, who invented printing with movable type around 1437.

Once all the elements of a technical document — text, illustrations and colour separation — have been brought to final artwork stage, ready for printing, the result may be reproduced by one of three different printing processes: letterpress (or relief printing); intaglio or gravure; or by offset lithography (a surface printing process).

The choice of the technique to be used in each particular case may depend upon the original artwork, the purpose which the job is to fulfil, the shape and size of the paper, the size of the print run (number of copies) and the quality of reproduction required.

For the most part, readers of this chapter whose technical documentation is produced by an external printing source, will almost certainly be using the offset lithography process. However, you may find it helpful to understand why this process is favoured, and in what ways it differs from other processes.

Strictly speaking, the decision about which process to use should be made before work on the documentation begins, since the nature of the layout of general print requirements largely depends upon the printing process used. Given however, that most technical documentation will be limited to standard sizes and formats, there is little point in discussing the details of design insofar as they relate to various printing processes. This should be a subject of separate study, should the nature of the publications work that you undertake become more varied.

As mentioned before, printing from cast letters was invented by Gutenberg. The composing of the type was always done by hand and sometimes still is, though more rarely now. In hand setting the text is set, line by line, on a 'composing stick', and then arranged consecutively by lines on a 'galley'. Proofs are 'pulled' off these galleys, often on a hand press, to provide copies for proof reading. After revision of the proofs, the text is assembled into pages according to a layout 'rough', together with illustrations.

The introduction of mechanised setting gave rise to two of the most familar names in typesetting, Linotype and Montotype. In the case of the Linotype machine, as its name suggests, each line of type would be cast as a single complete slug of metal. A similar process was also performed on a machine known as the Intertype. An alternative to this mechanised setting was the Monotype, which cast single characters and introduced the concept of 'movable type'. The principle employed consisted of the casting of type characters by means of a corresponding series of perforations on a paper reel. In both types of mechanised setting, the control was via a keyboard much like that of a typewriter.

The setting of type in the ways described above was exclusively for the oldest printing process of all, letterpress, which is the subject of the first of three printing processes described below.

LETTERPRESS

The letterpress printing process is one based on 'relief printing', and as such is the oldest form of printing. The block printing of the Chinese was a form of letterpress, even a 'John Bull' rubber stamp printing system could be described as letterpress. Generally today, however, letterpress is based on a photographic engraving system whereby printing 'plates' are made. In all cases, the printing area is raised for receiving the ink which is transferred to the paper by means of applying pressure (see Figure 12.1).

The letterpress process which is based on individual blocks, with illustrations separately engraved from the type, has an advantage even today. Because the type and illustration is built up in separate compartments and then composed to form the full page, individual components can be changed without the need to make up the entire page or making a new plate.

As mentioned, modern letterpress uses an engraving method for producing the etched areas to leave the relief surfaces, and the techniques employed are based on a photographic engraving method. This system of letterpress is called photoengravure, and the process of engraving plates is carried out by a person called an 'engraver'.

The making of a plate for letterpress involves bringing a photographic negative into contact with a light-sensitive metal plate. The exposed areas of the plate which are subjected to light passing through the negative, are light-hardened. The plate is then dipped into a bath of acid which etches to the required depth all areas of the plate which are not hardened. This leaves a relief to which the ink is applied.

The letterpress form of printing has been somewhat dramatically overtaken by 'offset' printing (see section on offset below), but this has not meant that letterpress is a poorer quality, outdated process. Letterpress has been traditionally used for printing newspapers, and has the advantage of being suitable for cheap materials and inks. The major disadvantage of letterpress is the time required to set up the printing plates, and it is in this area that offset printing has made considerable progress. There is no doubt that letterpress is capable of providing a high speed, high quality print process, and can maintain a consistent quality throughout the print run.

GRAVURE PRINTING

This process of printing dates back to the 15th century and is based

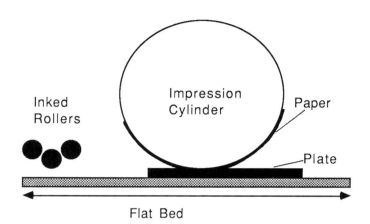

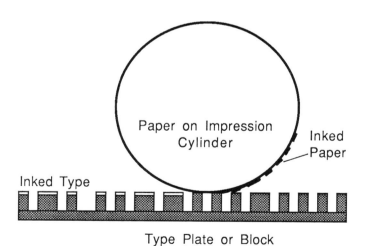

Figure 12.1   Letter Press — Flat Bed Process

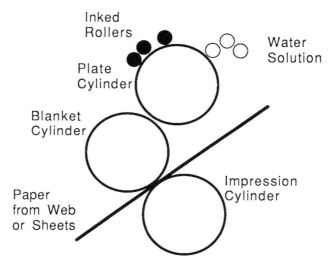

Figure 12.2 Offset Lithographic Process

on the form known as intaglio. The process works by cutting or etching the image to be reproduced into the surface of the printing plate. Ink is applied to the plate, and then residual ink wiped off the raised surfaces, leaving ink lying in the incisions of the plate. Using pressure during the print process, the image is transferred as the paper draws the ink from the incisions. This makes the difference between letterpress where it is the raised surface that carries the ink. Gravure plates can be made which may be wrapped around cylinders, or a cylinder can be engraved itself. In both cases, the engravure process is now carried out by a photographic process.

While in other printing processes it is usually only photographic artwork that has to be screened, in gravure printing all copy must be screened, including the text (see Chapter 5 for details about the screening process). The various tones of the process are controlled by the depth of the engraved dot, such that the deeper the engravure, the more ink is held, and hence the darker the image.

By nature of the way in which the plates are made, most gravure printing is carried out on rotary presses which can be fed from a continuous roll of paper (this is known as a web-fed process), or by individually fed sheets. Gravure printing, because it uses rotary presses, has the advantage of printing speed. It is therefore ideally suited for printing high quality, long print runs at a high speed. Consequently, it would not be a suitably economic way of printing technical documentation of low runs.

The plates themselves are also more expensive to produce than other methods of printing, but they have a longer life. However, if plates need to be altered or duplicated, this can be expensive. One of the advantageous characteristics of the gravure print process is that it can provide one of the richest blacks and widest tonal ranges compared to other processes, but this is unlikely to be relevant to technical documents, unless some clarity of illustration work is of paramount importance. The process does lend itself to a wide range of paper qualities and is at its best on smooth, coated papers, which are noticeably more expensive than uncoated stock.

OFFSET PRINTING (LITHOGRAPHY)

The offset printing process is based upon the principle of lithographic printing. Lithography is the process of printing from a flat stone. The image is applied to the surface with a greasy material and the surface is wetted. Greasy ink is applied and absorbed by the greasy parts of the surface but repelled by the wet parts. Prints can then be taken from the surface.

Offset lithography, the modern version of this process, uses thin, flexible metal plates which can be wrapped around a cylinder (see Figure 12.2). This makes them suitable for use on rotary presses, and is therefore now becoming more widely used, instead of letterpress, for the high print run requirements of newspaper printing. However, the process of platemaking is economical enough for it to be considered as a suitable method of printing most types of work.

The line and halftone film negatives, which have been taken from the final artwork, are assembled into what is called a 'flat'. Photographic paper proofs may be made of these flats and submitted to the client; these may be referred to as 'ozalid' proofs, or by various other names which reflect their general colour, for example: Van Dykes, blues, browns or silvers.

PRINTERS AND PRINTING

Printing plates are made from the flats by a photographic process. Plates can be made of one of a variety of materials, including paper (which is most economical, but only suitable for small print runs due it its short life), stainless steel or aluminium. The image is transferred from the flat to the plate by exposure to a high-intensity light and the plate is then developed, either manually or automatically, by a process similar to that used in photography. Plates can be made from either negative or positive films. The plate then has to be treated with chemicals that will reject the water solution and accept the ink (this being the lithographic process).

Offset lithographic printing can be used for a wide variety of jobs and account for most of the printing of technical documentation produced externally.

For most requirements the platemaking process is relatively inexpensive, compared to letterpress or gravure, and is faster to complete. Duplication and alteration of plates can therefore be achieved at a relatively low cost. The offset process can handle most types of paper and prints well on both rough surfaces and coated quality papers, and is particularly suitable for colour print processes. All presses used in offset are the rotary type, and may be fed either from continuous rolls of paper (web-offset) or with single sheets.

The term offset comes from the process by which the image on the inked plate is transferred, or 'offset', onto a blanket cylinder which in turn transfers the image to the paper. The reason the image is offset to a cylinder before being transferred to the paper, as opposed to being transferred direct, is to avoid damage to the fragile surface of the plate, which could be caused by the rough surface of the paper. The blanket cylinder, being made of rubber and therefore flexible, can mould to the texture of the paper, hence the ability of offset to print well on rough surface papers.

CONSIDERING COLOUR PRINTING

Colour printing involves additional expense, as separate plates need to be made for every colour applied and because colour presses are more expensive to use. Also, if colour photographs are used, for example, a process of scanning will be needed to separate the base colours. As well as these factors a better quality paper may need to be used to ensure good reproduction and the cost of colour proofs, usually called 'chromalins', will be significantly more than that for single colour jobs.

Technical documentation usually only goes as far as the application of a second colour, which is used to enhance the layout or highlight the text, for example, for picking out certain key words in the otherwise black text. This means that two sets of plates will be required. Therefore, if you are considering adopting a two colour format for your documentation, you are strongly advised to check 'ozalid' proofs before printing to ensure the correct colour separation.

Doubling the colours in your documentation does not double the price and, especially for large print runs, you may consider that the enhanced presentation gained by adding an extra colour is worth some additional expenditure.

13 Presentation and Packaging

Presentation and packaging of documentation is usually decided upon having considered the following:

a) Who will read the manual
b) In what market the documentation may be competing
c) The budget available
d) The print run
e) The method of revision and update

FORMAT

Information contained within Digitext's survey indicated that A4 format is still one of the most common for technical publications, with A5 size coming second. While the survey pointed out that this was surprising given that the 'industry standard' had long been regarded as A5, the reason for A4's popularity may have to do with the increase in DTP produced documentation. Desktop laser printers are, in general, designed to accommodate A4 photocopier paper and this may influence the decision on document format for the publishing department.

There are some specialist documents that use A3 size to incorporate, for example, circuit diagrams and not surprisingly are bulky to handle. The use of folding pages is probably more suitable, though this can increase the cost of collating and finishing.

COLOUR

Colour in technical documentation still seems to be little used in general, though it is quite widespread in the specific area of user documentation for software products. The use of colour in this area tends to be dictated by the presentation of competing documentation rather than as the result of a conscious decision about what the second colour will be used for. Whatever its use, a second colour will enhance the appearance of the documentation if used tastefully.

Most technical documentation jobs are in single colour (black) because the majority of publications have a print run of less than 250 copies. Colour work involves additional platemaking which is expensive and can therefore only be justified if the unit cost can be reduced adequately by a large print run. Use of selective colour pages may be possible, however, and now that photocopiers are becoming more sophisticated, with colour toner cartridge options and selective colour area copying, we may see low cost, photocopied documents showing a little more sparkle than before.

ILLUSTRATIONS

Graphics will enhance the presentation of the documentation. The type of graphics included will, like other factors regarding presentation, depend upon the cost factor. Photographic illustrations or colour plates will inevitably be more expensive than line illustrations. However, the use of a scanner and a DTP system will allow photographic illustrations to be incorporated with the text. This not only reduces the cost of the artwork, but dictates that all illustration material must either be line work or black and white scanned photos. Given time, all technical documentation departments will be using some form of integrated text and graphics processing systems, and the occurrence of graphics in technical documentation will certainly increase dramatically.

PAPER QUALITY AND METHOD OF PRINTING

With some 30% of technical documentation being produced by photocopying, there are limitations to how the presentation of the documents can be improved by the use of other materials. Those documents that are printed through an offset process can vary in quality depending upon the paper used.

PRESENTATION AND PACKAGING

There are many paper types to choose from and different paper surfaces will suit some types of printing better than others. Indeed if you want to go into the job in detail, the typeface may dictate the type of paper that should best be used. For example, a book face is generally enhanced by the 'squashing' of the ink during the print process, and this is best achieved on a rough paper. Sans serif faces are best presented on wood free, coated papers for clarity and so on. These are matters of a pedantic nature, but it is as well to at least consider some of the aspects of design that were employed when the typeface you have chosen was created. A book on type or typography may give you some help in this direction.

If the artwork is laser printed, the text will generally be much clearer than word processed copy, even if you photocopy the document. Again, costs will dictate what you can achieve in enhancing the presentation with quality papers.

BINDING

Since a good deal of technical documentation is photocopied, the binding possibilities will be dictated by this fact. Spiral binding seems to be a popular method of finishing documentation, though heat bound documents are becoming more popular with the availability of low cost desktop heat binding devices. Ring binding is by far the most popular form of finishing documentation and this can cater for both A5 and A4 documents alike, and allows for simple updating of individual pages that may be revised. If you spiral bind the documentation in-house, you still have the facility to include page changes and, by its nature, technical documentation is subject to change in fairly short timescales.

It is reasonable to assume that the method of binding adopted by most technical documentation publishers is governed more by the need to be able to update the documentation than the matter of cost. This is particularly true when you consider that the cost of ring binders and slipcases, popularly used for software documentation, is generally more expensive per unit that perfect binding or using hard bound covers.

The type of binding is also dictated by the need to be able to keep the documentation open flat on a surface for ease of reference, therefore perfect binding, stapling and even plastic spine binding can be regarded as a nuisance by the user who needs to refer to the documentation regularly.

IN CONCLUSION

The factors mentioned at the beginning of this chapter all affect the choice of presentation and packaging of documentation. However, it is interesting to note that in Digitext's survey, the majority of suggestions given in response to a question of how respondents thought that their documentation could be improved, said that it could be improved by better presentation and print quality, or other associated aspects such as use of colour, improved page layout, use of illustrations, etc.

This indicates that, although the general consensus of opinion as to the most important factors for good technical documentation are those of clarity, technical accuracy and readability, many technical documentation departments feel that presentation is generally of a lower standard than they would like to see.

14 Revisions and Updates

DESIGN FOR CHANGE

Almost all technical documentation, by the nature of the subject matter which it covers, requires updating. Documentation for technically developing products, which may include hardware, software, firmware and the like, may undergo significant alterations or enhancements. Just how to go about updating documentation efficiently and effectively requires close scrutiny.

There are many factors which will influence how you will go about the updating process. Economic factors will almost certainly be high on the list of priorities and also what time you have available. Ideally, documentation could be completely updated when a change is necessary, with a new issue completely replacing the old. However, this is rarely a realistic solution. For one thing, it is expensive, particularly if such a major change cannot be carefully co-ordinated so as to minimise the amount of stock of 'old' documentation that may need to be scrapped. Much will depend upon timing. If you have enough warning of changes that need to be made, you may schedule rewriting of documentation to coincide with the release, for example, of the new version of product to which it will relate. Nevertheless, this method of revision is certainly costly if the documentation changes frequently.

Documentation needs to be designed for change if changes are going to be made. The more frequent the changes that will be made, the more flexible must be the documentation to cope with many updates without detracting from its structure and ease of reading. There are many examples of technical manuals that are updated by the inclusion of a wad of pages containing 'Documentation for version X changes'. This wad of paper often bears no resemblance to the main documentation and provides no means of cross referencing with existing material. Consequently, the reader is expected to read the original manual first, then read the update information which may supersede what he or she has already read.

Updating of technical documentation has to be done for the convenience of the reader and not for the convenience of the publisher. Supplying pages of loose leaves which do not integrate with the existing manuals is certainly no help, although this method may be the only way of providing information in a short time scale, and it has to be said that under those circumstances alone, it is considerably preferable to supplying no new information at all. So, if the subject of your documentation changes suddenly and you need to supply new information very quickly, do not hesitate to send out something while you plan a more permanent update, but take a little trouble to make it useful.

If the change is very minor in terms of amount of words, think of ways in which you could provide that information so the reader can include it into what they already have. For example, if a single paragraph is affected due to some change in technical specification, what about using a self adhesive label with the new information on it that the reader can stick over the old text? By providing the detail this way, the reader can at least keep the new item in the relevant place when reference to it needs to be made, and it's quite permanent — not easily mislaid without the original page going missing too.

METHODS OF PRESENTING UPDATES/REVISIONS

Many technical publications are presented in ring binders. As discussed in Chapter 13 on presentation and packaging, this usually has two major benefits: it is easy to lay the documentation flat on a surface for ease of reference; but more significantly, it is easy to update in a cost effective way. By supplying new written material on pages drilled to fit the ring binder, new information can easily be added, and out-of-date pages may be removed and replaced with the new. This method of updating and revising documentation remains a

REVISIONS AND UPDATES

favourite and certainly helps to maintain the existing structure of the original format and makes it easier to keep changes in their relevant place. It does, however, become an untidy means of updating large numbers of pages and can be difficult to administer.

Update pages have to be referenced. You should adopt a rule of including an issue number and date somewhere on the page, perhaps as a footer. This need not be very prominent provided it shows that a certain page has been updated. This means of identification will help should any queries arise about how up-to-date an individual's documentation is. There will invariably be those who do not receive an issue of update pages, and when they query their documentation, it is a simple matter to check what level of documentation they have by reference to appropriate pages that should be so marked.

Unless you intend to update your documentation very regularly, a month and year will suffice as the reference date preceded by an incrementing issue number. Some technical documentation accompanies products which themselves have version or issue levels and are referenced by some number or letter. For instance, software often carries some version or ship level number, so that it is clear what the latest product is. Under these circumstances, it is preferable if you can tie in the issue reference on update pages with the issue level of the product. For example XYZ Software Version 3.00 is accompanied by documentation level 3.00. Any pages which have been supplied exclusively to cover the features of the version 3.00 product can be identified with the words 'Issue 3.00' and the date of their first issue.

If you are supplying many update pages it can be helpful if you provide a covering page or pages which identify the pages that have changed. This serves as a check for the recipient that all the pages which are due to be updated have been received. If you have time, and it doesn't really take too long, you should include a brief note to cover the nature of the change. This simply saves the reader from having to scan through all the update pages to see what has changed. This is particularly appropriate where the update pages cover enhancements to an existing product manual.

Only some of the changes may be relevant to the user, and while you expect them to replace all the pages supplied, do not waste their time by expecting them to read something which has no importance to them. These notes need only be very brief. Don't repeat the work you've done in the update pages themselves. For example, use brief notes in a format thus:

Replace Pages:	With Pages:	Because:
12	12	Text error in para. 3
22	22	New switch option documented

and so on.

The heading 'With Pages' may seem pointless in the above example, since it is unlikely that you would want to change, say, page 12 with any other page. But you may adopt a method whereby, if so much new information is required that it cannot be accommodated just by the replacement of the existing page, you could include sub-pages to cater for the overflow of information. In this case you could use a folio with the suffix a, b, c, etc for the extra pages. The beauty of this approach is that it can save you having to alter the subsequent page numbers, or repaginate whole chapters. So when page 12 changes to pages 12 and 12a and 12b, this is made clear on your accompanying sheet by the following reference:

Replace Pages:	With Pages:	Because:
12	12, 12a, 12b	Text error in para. 3
22	22	New switch option documented

Now it is clear that the original page 12 is replaced and two new pages are to be inserted, so the recipient should check for three pages to cover this update.

Some documentation uses what is called duo-decimal page numbering. This is a deliberate form of planning for change. In this case, a page folio follows the chapter or section numbering first, then a page number for that section only, so chapter 1 will begin with page 1.1, followed by pages 1.2, 1.3, 1.4 and so on until the end of the chapter. Chapter 2 then begins with page 2.1, followed by pages 2.2, 2.3, 2.4 and so on. The purpose is clear. If new pages need to be added to chapter 1, only the folio references of chapter 1 are affected. All subsequent chapters remain unaffected and consequently only the index for items in chapter 1 is affected. There are even some technical manuals which go to a third level, such that chapter 1 will

REVISIONS AND UPDATING

consist of sections 1 to 5. The first page of chapter 1, section one has the folio 1.1.1. The third page of chapter 1, section 4 has the folio page 1.4.3 and so on. Quite frankly, two levels is quite sufficient. Three levels of decimal page numbering is simply ghastly and is as bad as multiple sub-paragraph references.

It is up to you to choose the method which best suits your purposes. For my part, the use of page update suffixes such as a, b, c for pages which are conventionally numbered in sequence is by far the most reasonable means and has some dignity about it. Duo-decimal page numbering should be reserved for reference material such as lists of programming language commands or the like.

If you bind your own documentation (perhaps you have a spiral binder or use a plastic spine and covers), then it will be relatively simple for you to collate the new material into the documentation before it's despatched to the customer. This has the benefit of it seeming to be always complete, and the fact that it has been updated is virtually undetectable. It also makes sure that all information is kept together and there is no risk of losing new information unless you lose the lot. You may still have to consider how to number new pages though, unless you want to repaginate the entire document, and this point is covered later in this chapter.

This has one disadvantage however, a user of a previous version of the documentation who wants the latest information always has to be given a complete new set. This is okay if the documentation is sparse, but for a large volume or several volumes, this can be unnecessarily expensive and wasteful. If you have to supply the pages uncollated, you should consider how the reader can efficiently incorporate them into what they already have. This can be difficult if their documentation is spiral bound (you try unpicking a spiral bound manual and rebinding it!), so you may need to reconsider the design and presentation of the original documentation.

If your budget stretches to it, there is no better way than to supply newly bound up-to-date technical documentation whenever it has been updated. From the reader's point of view, this approach will maintain a prestige image for your company. Let's face it, if you've just paid out £XX for a new product version you would not be pleased to receive a bunch of photocopied update notes for an otherwise well produced, high quality document.

Whatever standard you set in the presentation of your original document, you should expect to maintain it in your updates and revisions. If you fail in this, the consumer will take the view that they have been hard done by, or that you care less for their interests once you have made the initial sale and provided them with the first release.

15 On-line Documentation

We are currently nowhere near the paperless office concept that the computer industry would have us believe exists. There are those who feel that printed documentation will be replaced by on-line documentation (instructions on a computer screen, for example) in due course. While this may be ultimately true, there is little sign of it yet. Most on-line documentation available is provided to accompany printed documentation and as such complements it rather than replaces it.

It can only be used where the facilities to show text on a screen are present, and so is restricted for the most part to computer applications. Help text for software packages is one common form of on-line documentation. Others include training diskettes, 'rolling software demos' (programs that perform a series of tasks automatically to demonstrate the features of a system), text files on disk for information such as installation guides, and installation routines themselves.

More advanced on-line systems may include interactive documentation which can be called on screen at the appropriate time, providing help which is relevant to the process currently being performed. This is known as 'context sensitive' documentation. For the latest in on-line documentation, expert systems are now being employed to provide help. An expert system is an artificial intelligence software application that effectively provides knowledge about a subject, much like a human expert. The artificial expert can provide advice or answers to queries, or guidance based on a sequence of instructions or processes. The expert system can query the user for details, or the user may volunteer information to the system, on which judgements and responses can be based. More recently, video based training aids are being used for on-line training.

All on-line documentation must, to a certain extent, be given similar considerations to content and design as printed documentation. You need to know the audience that will use it, and provide the most appropriate information in a concise and easy to use way. One area in which on-line documentation comes into its own is for documenting rapidly changing information. You may be familiar with software applications that include a 'READ.ME' file on disk, which includes the latest details on the version of the software, and you have to type the file to screen or output to a printer for reading. Since this is a file that accompanies the software application itself, it can carry the latest details without the need for worrying about printing timescales or lead times.

Installation routines for software packages are commonly handled with on-line documentation, since the user may have to provide certain information about the type of system on which the software is to be used. In this case, the installation routine itself may not be documented in detail within the printed manual, which can save on paper and print costs. The manual may simply refer the user to the installation file, and then the file itself contains the detailed documentation.

The decision to use on-line documentation may be based on a number of factors. For example, whether a printed form is actually needed, which form of documentation is most practical in the circumstances, whether graphics are necessary (which may be difficult to provide on screen), how quickly the documentation needs changing and so on. It is usually an ancillary task of the technical documentation department to design and maintain on-line documentation, since the technical authors are usually already involved in the process of providing the same information in printed form.

There are restrictions on designing on-line documentation compared to printed forms. You will normally have limited facilities for graphics, unless you are documenting a system that already uses

ON-LINE DOCUMENTATION

high resolution screens (eg on-line help for DTP systems can be very effective), and the choice of typestyles may be limited to just upper and lower case character styles with perhaps some underlining.

When providing on-line documentation which may be accompanying printed documentation, it can be useful to the user if you cross refer to the printed form from the on-screen version. Therefore, when a user finds that the normally limited information provided on-screen is not enough to enable a comprehensive understanding of the subject or concept, they can be referred to a more extensive set of instructions within the printed documentation.

Generally, on-line documentation is briefer and simpler than the printed form. This may be dictated by the memory or workspace available for displaying help or documentation on-screen whilst an application is loaded.

As a final note, you should not provide an entire manual on disk and expect the user to print this out before using the software. This is an annoying practise adopted by some software manufacturers and often employed by the vendors of cheap, off-the-shelf software. It is not recommended as a satisfactory way of providing technical documentation to a user purchasing a hardware or software package to be used in a business environment.

16 Development

Part of the duties of publications managers and authors alike is to endeavour to develop the documentation and the documentation process for which they are responsible, in order to bring about improvement. It is unlikely that any technical document is as good as it can be and there may be restrictions on development that are outside your control. For instance, if you are given a restrictive budget to work to, you may be unable to make changes in the methods employed to produce your documentation.

Previously, introducing technology into the process has not been cheap, though desktop publishing is responsible for providing the means to improve the quality and appearance of technical documentation at an affordable price. This will only work in your favour, however, once you have taken the trouble to understand the principles of DTP and how to make best use of it. As education in this area improves, we should expect to see better quality technical documentation across the computer industry. However, the authors and publications managers have a duty to the industry to ensure that, by utilising this new technology, standards of quality are maintained and improved. This means directing efforts towards training and gaining knowledge about typography and print processes, about good writing skills and design and presentation attributes.

The training and development of authors under your control, where applicable, is an important part of your department's development. While there is little training available that specialises in training technical authors for the industry, it is up to the technical publishers to take steps to put that right. Phoenix Technical Publications Limited instigated a series of meetings for publications managers, which has resulted in bringing together the interests of people involved in technical publishing for the computer industry and attempting to establish some common ground between them. By doing this, not only can we share in the problems that the job of producing technical publications brings about, but work together to solve those problems.

Improving the quality and presentation of documentation requires a conscious effort. So often one hears technical writers defending their publications rather than constructively criticising them. It is natural to think that your own work is as good as it can be under the circumstances and that the restrictions on improvement are down to lack of resource.

I have already mentioned that, according to Digitext's survey, the main points of action that respondents to the survey gave for improving their documentation were in the areas of presentation and printing. Only two respondents suggested that attention to style and content might improve their documentation, and two others that more accurate technical detail was required. This suggests that the majority feel that their restrictions in the field of technical publishing are financially based. However, increasing resources will only partly assist in relieving these restrictions.

Development towards a world of better documentation must focus on ways in which the difficulties of communicating ideas and information to readers can be overcome. In this respect, many technical publications have yet to overcome one major problem, that of getting their readers to use the documentation in the first place. Documentation often only gets attention when the user of a technical product has failed to obtain information from another source, which they had thought to be easier to comprehend. This may mean learning by trial and error, asking questions of others nearby, or phoning the supplier, etc. Also, documentation which is bulky can put off users, who do not feel they can afford the time to read it. For the non-technical audience, any aspect of the documentation which 'gives away' the fact that it contains 'technical' information can create a psychological barrier.

DEVELOPMENT

Perhaps the answer lies in communicating technical information by means of graphics, or perhaps you believe that technical documentation in printed form will not provide the answer to the development of better communication, but that this lies in the field of on-line computer based training and documentation. Above all, you should take the trouble to at least consider ways in which you can develop the technical documentation standards of the future, rather than wait for them to impose themselves on you. In short, the development of technical documentation lies in your hands.

Appendix A
Contractors

Further information about the use of contractors or documentation houses can be found in Chapter 2. Below are details of just a few major operators in this field, which will enable you to establish some initial contact should you require documentation writing services to assist your workload, or to overcome lack of expertise in your own organisation.

Digitext Limited
15 High Street
Thame
Oxon OX9 2BZ

Telephone: (084 421) 3434

Digitext offer a service for all categories of contract documentation staff: publications managers, technical writers, senior technical writers, editors and marketing copywriters.

They can help you meet your documentation requirements — from first-line publicity brochures to complicated technical literature. The products Digitext have documented include databases, communications software, spreadsheets, word processors, languages, operating systems, etc, for micros, minis and mainframes.

Phoenix Technical Publications Limited
Barford House
Shute End
Wokingham
Berks RG11 1BJ

Telephone: (0734) 774211

Contact: John Gardner

Phoenix's writing skills extend from copywriting and journalism to technical authorship. This enables them to apply a consistent approach to a range of documentation. For long and complex tasks, Phoenix have experienced project managers and publications consultants to assist your organisation. They set targets, monitor progress, maintain quality and liaise with outside services such as designers and printers.

K3 Software Services Limited
Severn House
10 The Moors
Worcester WR1 3EE

Telephone; (0905) 29821

Telex: 336700

Contact: Stuart McPhee

K3 have a documentation service where a team of technical writers are available to satisfy both technical and end-user documentation requirements. Their services are offered in three main areas:

a) Documentation consultancy (including development of strategies, training, standards development, etc);

b) Retrospective systems documentation, eg from source code (documentation of existing systems which may never have been initially documented, and knowledge of the system may be with staff who have left, etc);

c) Traditional end-user documentation writing.

CONTRACTORS

TMS Computer Authors Limited
The Sheilings
The Street
Wonersh
Guildford
Surrey GU5 0PE

Telephone: (0483) 898606

Contact: David Preece

UK staff: 35

Main Business Areas

TMS Computer Authors is a documentation house which provides top level documentation and training to computer manufacturers, software suppliers and users in industry and commerce.

They have the experience and qualifications to undertake documentation and training assignments, ranging from the preparation of simple user guides through to the provision of all documentation and training for large scale development projects; and have the technical and managerial skills to take responsibility for major documentation projects from feasibility right through to production.

Services and Products

They provide documentation and training consultancy and undertake feasibility studies, planning activities and project management for documentation and training projects. They prepare user documentation, systems documentation, and training materials, and increasingly they are being called upon to prepare business procedural documentation.

They are used to structuring documentation suites, providing standards for documentation and setting procedures for documentation maintenance. They provide a design activity from simple page layout through to full colour work for sales literature.

The majority of their work is managed by a TMS project manager, although they can also provide technical authors on contract. They offer a specialist recruitment facility for client companies.

Vertical Market Specialisations

All systems and software developers and producers, systems and software users; finance, particulary banking and securities.

General Company Description

TMS was established in 1982 and now provides consultancy, documentation and training services to many of the UK's major information technology suppliers and users.

TMS offices in Guildford are close to areas of major activity in London and the Thames Valley and ideal for access to other parts of the UK and Europe.

Activities Overseas

Documentation and training, technical authors on contract.

Graded Business Importance

1 Documentation house
2 Consultancy
3 Education and training
4 Recruitment

Location of Head Office

South East

Appendix B
Author Recruitment

The subject of author recruitment has been mentioned in this book in Chapter 9 on the role of the publications manager. As with any other profession, there is assistance available to help you recruit individuals for vacancies in this field. Most organisations that provide services for recruitment do so for a wider range of job types. For example, there are specialist recruitment agencies who deal solely in selection of computer industry candidates. There are a few who concentrate on programmer/analysts. However, to find a recruitment service specifically for authors, you will need to look to the documentation houses and technical authorship organisations like Phoenix and Digitext. Their details are repeated on the following pages, with details on their recruitment facilities.

DIGITEXT DOCUMENTATION STAFF RECRUITMENT SERVICE

Digitext offers a complete recruitment service for all categories of documentation staff: publications managers, technical writers, senior technical writers, editors and marketing copywriters.

Digitext will discuss with you the responsibilities of a particular job, the available career path, the experience and qualities required of candidates, the salary range and conditions of employment. They then prepare a job specification, search their files for suitable candidates and/or can prepare an advertising programme, and handle interviews of the most likely candidates. Final interviewing and selection is then made by yourselves.

For further details contact:

**Digitext Limited
15 High Street
Thame
Oxon OX9 2BZ**

Telephone: (084 421) 3434

Contact: Bob Ritchie

PHOENIX TECHNICAL PUBLICATIONS RECRUITMENT SERVICE

Phoenix's author recruitment service is simple and efficient. They are principally a documentation house and have expert knowledge of technical authorship. Two of their directors have many years major company experience of finding and selecting suitable candidates for technical author positions. They have been highly successful in this field and can offer a workable solution to your recruitment requirements, whether for an individual or a number of authors.

For further details on the recruitment service, contact John Gardner at:

**Phoenix Technical Publications Limited
Barford House
Shute End
Wokingham
Berkshire RG11 1BJ**

Telephone: (0734) 774211

Appendix C
Author Training

There is little academic training available for technical writers that specialises in the subject of authorship. This makes the selection of author candidates for recruitment difficult to base on a recognised qualification. There are a couple of courses that result in a certificate or diploma, but other training is seminar based, and is independently arranged by commercial organisations. These can be helpful, however, for those of a technical background who are already involved in documentation or technical writing, and need a little extra guidance or steering in the right direction.

GLOSCAT DIPLOMA IN SOFTWARE DOCUMENTATION

Gloucester College of Arts and Technology have, for a number of years, run a course for industrially/commercially experienced recruits leading to a Diploma in Software Documentation. This is a one year full-time advanced course, for students at a professional level. It is designed to equip them for employment as technical authors/software documentors in industries that make, supply or use computer systems.

Course Objectives:
The course aims to equip people of a high academic standard who have industrial or commercial experience, preferably at a professional level, for employment as technical authors in the computer industry. On completion of the course, students will be able to plan, organise and write technical literature from original sources

for users and operators of computer systems. Students will also be able to document clerical procedures, and prepare and maintain systems documentation and other data processing documentation. They will be introduced to the main functions carried out by systems analysts and high-level language computer programmers, and will be able to undertake relatively simple tasks in both of these fields. Students will become conversant with hardware and software, and will be able to use technical vocabulary correctly and fluently.

Course Structure:

For the first 26 weeks of the course, the students are college based. During the last ten weeks of the course, they work as technical authors/software documentors in companies and organisations which agree to provide this industrial experience. The students' course fees and personal allowances are paid by the Manpower Services Commission's Training Opportunities Scheme (TOPS) while the students are both college based and on their industrial placement.

Course Content:

The following major topics are covered in the course structure:

- Technical Communication
- Computer Systems and Documentation
- Program Writing
- Microcomputers and Data Communications
- Science and Technology
- The Electronic Office

Assessment:

To complete the course successfully, students must satisfy the Course Committee that they have reached the required standards in:

1) Coursework — students' progress on the course is continuously assessed by individual and group assignments. These draw on information and skills learnt in different topic areas and ensure the integration of course material.

    All students are required to complete a substantial documentation project during the college-based part of the course.

AUTHOR TRAINING

This documentation must be of a professional standard and forms the basis of verbal presentation to an audience of staff and students.

Whilst on their industrial placements, students will be assessed by their supervisor within the placement organisation and their industrial tutor from the college. A satisfactory report is required before the Diploma can be awarded.

2) Examinations — at the end of the college based part of the course, students sit three-hour examinations covering the course material.

PHOENIX TECHNICAL PUBLICATIONS LIMITED

Phoenix Technical Publications Limited offer a series of short courses on technical authorship, report writing and communication skills. Their courses last from one to five days and are either public or tailored to the needs of a particular client. Phoenix course tutors have worked professionally in computing, publications management, technical journalism and education. It is this mixture of practical and theoretical experience that enables them to offer a constructive training programme:

Writing Documentation for Computer Users (Introduction)

A five-day course for junior technical authors covering basic documentation principles.

Writing Documentation for Computer Users (Advanced)

A three-day course for authors with one to two years experience. It covers the documentation cycle from initial planning to final evaluation and includes writing for international audiences.

Documentation Project Management

A three-day course for experienced authors who are anticipating a move into team leadership or publications management.

Report Writing

A one-day course (including a half-day writing seminar) intended for anyone who has to write business or technical reports but who has had little training in writing.

Effective Technical Writing

A three-day course for programmers, systems designers and others who have to write technical documents but whose main job is not authorship.

Effective Business Writing

A two-day course for anyone who has to write business reports, proposals, memoranda or letters for an internal or external readership. It also includes writing for international audiences.

Effective Presentation

A three-day course for anyone who has to convey information verbally to large or small groups.

For further details contact:

Keith Mason
Phoenix Technical Publications Limited
Barford House
Shute End
Wokingham
Berkshire RG11 1BJ

Telephone: (0734) 774211

THE INFOMATICS RESOURCE CENTRE

1) How to Design On-line User Documentation

Purpose:

A two-day workshop designed for systems analysts, programmers, information systems managers, technical writers and others in the computer field who are currently developing or planning to develop on-line computer documentation. Participants will examine and practise the process of developing various forms of on-line documentation, from tutorials to help systems. At the close of the workshop, participants will have a standard set of guidelines and techniques for planning and developing on-line documentation.

Outline:

Introduction to On-line Documentation
The Life Cycle of an On-line Documentation Project

AUTHOR TRAINING

How to Make On-line Documentation Visually Appealing
Writing and Designing On-line Menus
Writing and Designing On-line Assistance
Writing and Designing On-line Tutorials
Available On-line Documentation Systems
Future Trends in On-line Documentation

2) How to Write and Produce Effective User Documentation

Purpose:
This two-day seminar is designed to enable the delegate to avoid the common pitfalls of user documentation by employing writing techniques, text formats and effective graphics which will make your material more 'user friendly'. Step-by-step guidance and practice sessions show delegates how to plan, structure, write and edit material for maximum clarity and precision. Important consideration is also given to techniques for revising and controlling documentation that has already been produced.

Outline:
Introduction
Planning
Writing a Procedure
Choosing Text Formats
Making the Information Accessible
An Introduction to On-line Documentation
Drafting the Manual
Editing and Publishing
Wrapping Up
Controlling and Revising

For further details, contact:

**The Infomatics Resource Centre
2 The Chapel
Royal Victoria Patriotic Building
Fitzhugh Grove
London SW18 2SX**

**Telephone: 01-871 2456
Telex: 299 180 MONINT G
Fax: 01-871 3866**

Appendix D
Copy Editing and Proof Correction

When editing and amending typescript copies or printers' proofs, it is a good idea to get into the habit of using a standard method of correction marking. This will help clarify the corrections to be made, and if the mark up symbols used are universally recognised amongst the parties involved, it will save ambiguities and time wasted in incorrect amendments.

Some of the most common forms of copy correction marks are shown on the following pages. They are extracted from BS 5261: Part 2: 1976, and are reproduced by permission of the British Standards Institution. Complete copies of the document (of some 16 pages) can be obtained from BSI at Linford Wood, Milton Keynes, MK14 6LE.

If you have come across information on the BSI standards of mark up, you may have noticed that the correction of proofs should be done in an appropriate colour of ink, depending upon who is marking the error. This tends to be only relevant when you are trying to clearly define who will be charged for the corresponding amendments, and is more appropriate to book publishing than it is to technical documentation production. Generally, you should aim to make your corrections clearly visible in a different colour to the copy.

A corresponding mark in the margin will highlight an error, which may otherwise be missed if it is only marked within the text. If you use red ink on black copy the corrections will show clearly. If you are giving a proof of a document to someone to check, tell them to mark errors clearly in ink. I have noticed, through past experience, that people not used to correcting proofs tend to be afraid of spoiling them (particularly typeset page proofs) and consequently mark corrections lightly in pencil. Errors marked up in this way can easily be missed.

Using a standard form of copy correction marking becomes more important when you mark proofs for printers and typesetters. They may be used to the BSI requirements, so whatever system you adopt for internal use may have to conform to their peculiarities if an error correction is to be made clearly understood. I witnessed a proof of a colour brochure with errors marked, not for printer corrections, but for query internally where some words to be queried were underlined. Unfortunately, the proof was passed back to the typesetter who reset the text and changed all the underlined words to italics. This misunderstanding illustrates how such ambiguities can be both costly and time consuming.

COPY EDITING AND PROOF CORRECTION 187

		In text	In margin
To substitute		ma~~k~~e	d
	OR	they ~~tend~~ letters	give
To transpose		li̶t̶	list
	OR	to|boldly|go	⊔⊓
To delete		pur~~g~~e	⌢⌒⌢
	OR	he also cared ~~too~~	⌢⌒
To insert		ti⋏t	gh
	OR	they/gone	had
To close up		ove͡r reach	⌢⌣
To insert space		selfknowledge	Y
To change to italic		<u>self</u> knowledge	⏛
To change italic to upright type		(*self*)knowledge	⊥⊥
To change to bold		the vector r̰	∿
To change capital to lower case		the Ⓢtate	≢
To change lower case to capital		the s̲t̲a̲t̲e	≡
To start a new paragraph		ends here.⌐In the next	⌐
	OR	the discussion ends here. ⌐In the next lecture Johns	⌐

	In text	In margin
To run on	ends here.⤴︎ ⤶In the next lecture	⤵
OR	the discussion ends here. ⊏In the next lecture	⊐
To insert space between lines	⟩the discussion ends here. ⟨In the next lecture Johns	
To close up space between lines	the cavalry ⎛ the paratroops ⎝ the gunners	
To substitute or insert note indicator or superior	According to Johns⁁this	²⁄
To substitute or insert inferior	the formula H⁄O is	⁄₂
To stet (if you make a mistake and want to restore the original)	the ~~saints~~ were important ~~bishops~~	⊘

To make punctuation changes:

Text mark / to substitute   OR ⋀ to insert

Margin mark	⊙	⊙⊙	;	,	ʼʼ	( )	⊢⊣	⊢	⊘
	full stop	colon	semi-colon	comma	quotation marks	parentheses	hyphen	dash	oblique stroke

Put / after each correction that does not already end in a caret (omission sign). This is especially important when two or more correction marks are required on one line.

Example: d / ⊙ / ≡ /

Appendix E
DTP Packages

The following pages contain entries submitted by some suppliers of DTP systems, listed alphabetically by supplier name. It should be emphasised that the list is by no means comprehensive, and is subject to change without prior notice by either the vendors, or the publishers of this book. The reference material provides an initial source for you to make some enquiries, and you should check the trade press for latest information and other suppliers.

Supplier Name and Address:	Apple Computer (UK) Ltd Eastman Way Hemel Hempstead Hertfordshire HP2 7HQ
Telephone Number:	0442 60244
Product Name:	Apple Desktop Publishing Systems
Approx Retail Price:	£4,000-£12,000 Complete System
Hardware Supported:	Macintosh
Operating Environment:	Mac O/S
Page or Document Description Language Supported:	PostScript, HP Laserjet, Diablo 630
Printers Supported:	LaserWriter IISC, LaserWriter IINT, LaserWriter INTX
Typesetters Supported:	Linotronic, Monotype, Compugraphic, AM Vastypen

DTP PACKAGES

Supplier Name and Address: Appropriate Technology
APTEC House
South Bank Business Centre
Ponton Road
London SW8 5AT

Telephone Number: 01 627 1000

Product Name: Gem Desktop Publisher

Approx Retail Price: —

Hardware Supported: IBM PC/AT/XT and Compatibles

Operating Environment: DOS 2.x or higher or DOS plus

Page or Document Description Language Supported: PostScript

Printers Supported: LaserWriter/HP Laserjet+/ IBM Proprinter/Epson, Dot Matrix 4020 and others

Typesetters Supported: —

Supplier Name and Address: Appropriate Technology
APTEC House
South Bank Business Centre
Ponton Road
London SW8 5AT

Telephone Number: 01 627 1000

Product Name: Aldus PC Pagemaker

Approx Retail Price: —

Hardware Supported: IBM AT Compatibles with 640K and 10MB

Operating Environment: MS-DOS/PC-DOS 3.0 or above

Page or Document Description Language Supported: PostScript

Printers Supported: PostScript/HPLJ+/IBM Page Printer 3812/and others

Typesetters Supported: Linotronic

Supplier Name and Address:	Appropriate Technology APTEC House South Bank Business Centre Ponton Road London SW8 5AT
Telephone Number:	01 627 1000
Product Name:	Z-SOFT Publishers Paintbrush
Approx Retail Price:	—
Hardware Supported:	PC/XT/AT Compatibles
Operating Environment:	DOS 2.0 or higher/512K RAM
Page or Document Description Language Supported:	PostScript
Printers Supported:	PostScript/HPLJ+/Epson/IBM Proprinter/JLaser and others
Typesetters Supported:	—

Supplier Name and Address:	Appropriate Technology APTEC House South Bank Business Centre Ponton Road London SW8 4AT
Telephone Number:	01 627 1000
Product Name:	Rank Xerox Ventura Publisher
Approx Retail Price:	—
Hardware Supported:	All PC/XT/AT/386 Compatibles with 640K and hard disk
Operating Environment:	MS-DOS/PC-DOS/DOS Plus/Concurrent DOS 2.1 or above
Page or Document Description Language Supported:	PostScript/Interpress/DDl
Printers Supported:	IBM Personal Page Printer/Epson/IBM Proprinter/HPLJ+/JLaser/Turbolaser/Xerox 4020
Typesetters Supported:	Linotronic 100/202/200/Compugraphic/Cora/Ace

DTP PACKAGES

Supplier Name and Address:	AST Europe Ltd AST House 2 Goat Wharf Brentford Middlesex TW8 0BA
Telephone Number:	01 568 4350
Product Name:	AST Turboscan
Approx Retail Price:	£1,295
Hardware Supported:	IBM PC and Compatibles and Macintosh
Operating Environment:	—
Printers Supported:	EYESTAR
Typesetters Supported:	—

Supplier Name and Address: AST Europe Ltd
AST House
2 Goat Wharf
Brentford
Middlesex TW8 0BA

Telephone Number: 01 568 4350

Product Name: Premiums 286, with 40MB hard disk

Approx Retail Price: £2,775

Hardware Supported: IBM Compatible, Lomitz, 0 wait state

Operating Environment: DOS 3.2

Page or Document Description Language Supported: PostScript Version 47

Printers Supported: Turbolaser/PS £3,695

Typesetters Supported: —

DTP PACKAGES

Supplier Name and Address: Canon (UK) Ltd
TDP Division
Canon House
2 Manor Road
Wallington
Surrey SM6 0AJ

Telephone Number: 01 773 3173

Product Name: Canon Express Desktop Publisher

Approx Retail Price: £8,950

Hardware Supported: Laser Beam Printer/IX12 Scanner, Canon A200EX Computer (40MB Hard Disk) IX12F Flatbed Scanner A2003 High Resolution Screen

Operating Environment: —

Page or Document Description Language Supported: 'Expression'

Printers Supported: LBP-SX and LBP-8 MK2

Typesetters Supported: N/A

Supplier Name and Address: Cognita Software Ltd
42 Ewald Road
London SW6 3ND

Telephone Number: 01 736 3637

Product Name: NEWSWRITER

Approx Retail Price: £666.66 (Single-User),
£1333.33 (Multi-User)

Hardware Supported: IBM and Compatibles

Operating Environment: DOS and Xenix

Page or Document Description Language Supported: —

Printers Supported: Canon LBPA1 + 2

DTP PACKAGES

Supplier Name and Address: DPS Typecraft Ltd
Hamilton Road
Slough
Berkshire SL1 4QY

Telephone Number: 0753 35156

Product Name: XTRASET

Approx Retail Price: £1,995-£3,850

Hardware Supported: IBM Compatibles

Operating Environment: PC-DOS

Page or Document Description Language Supported: Most Phototypesetters and PostScript

Printers Supported: Any IBM Compatible

Typesetters Supported: Autologic, Compugraphic, Linotype, Monotype, Varityper, Xenotron

Supplier Name and Address: Electronic Printing Systems Ltd
Katana House
Fort Fareham
Newgate Lane
Fareham
Hampshire PO14 1AH

Telephone Number: 0329 221121

Product Name: JETSETTER

Approx Retail Price: £295

Hardware Supported: IBM AT's/XT/PC or Compatible

Operating Environment: MS-DOS 2 upwards

Page or Document Description Language Supported: None

Printers Supported: HP Laserjet and/or Compatible

Typesetters Supported: None

DTP PACKAGES

Supplier Name and Address: Electronic Printing Systems Ltd
Katana House
Fort Fareham
Newgate Lane
Fareham
Hampshire PO14 1AH

Telephone Number: 0329 221121

Product Name: VENTURA

Approx Retail Price: £795

Hardware Supported: IBM AT or Compatible

Operating Environment: MS-DOS

Page or Document Description Language Supported: PostScript

Printers Supported: HP Laserjet Family

Typesetters Supported: None

Supplier Name and Address: Gestetner Ltd
Gestetner House
210 Euston Road
London NW1 2DA

Telephone Number: 01 387 7021

Product Name: Gestetner Desktop Publishing

Approx Retail Price: £8K Upwards

Hardware Supported: Apple

Operating Environment: Apple

Page or Document Description Language Supported: PostScript

Printers Supported: All PostScript Laser

Typesetters Supported: All PostScript or similar

DTP PACKAGES

Supplier Name and Address: Hewlett-Packard Ltd
Miller House
The Ring
Bracknell
Berkshire RG12 1XN

Telephone Number: 0344 424898

Product Name: The Hewlett-Packard DTP Solution

Approx Retail Price: From £7,500 plus

Hardware Supported: HP Vectra PC, HP LaserJetII Printer, HP ScanJet Desktop Scanner

Operating Environment: MS-DOS and MS Windows

Page or Document Description Language Supported: Aldus PageMaker and PostScript

Printers Supported: HP LaserJet Series II

Typesetters Supported: All PostScript compatible typesetters

Supplier Name and Address: Heyden & Son Ltd
Spectrum House
Hillview Gardens
London NW4 2JQ

Telephone Number: 01 203 5171

Product Name: Quark XPress

Approx Retail Price: £695

Hardware Supported: Apple Macintosh (Plus, SE and II)

Operating Environment: Apple Macintosh (Plus, SE and II)

Page or Document Description Language Supported: PostScript

Printers Supported: Any PostScript

Typesetters Supported: Any PostScript, Linotype

DTP PACKAGES

Supplier Name and Address: ISG Data Sales Ltd
Wellington Industrial Estate
Basingstoke Road
Spencers Wood
Reading
Berkshire

Telephone Number: 0734 884666

Product Name: Ventura/Pagemaker

Approx Retail Price: £600-£800

Hardware Supported: IBM Compatible

Operating Environment: MS-DOS

Page or Document Description Language Supported: PostScript

Printers Supported: H.P. LaserJet, PostScript, Kyocera, etc.

Typesetters Supported: Any PostScript Compatible

Supplier Name and Address:	Mass Mitec 52B London Road Oadby Leicester LE2 5DN
Telephone Number:	0533 718031
Product Name:	Ventura Publisher
Approx Retail Price:	£795
Hardware Supported:	Standard IBM PC/XT/AT or Compatible 10MB min
Operating Environment:	PC/MS-DOS Version 2.1 or above (512K) 3.1 version or above (640K)
Page or Document Description Language Supported:	PostScript
Printers Supported:	Xerox 404s, 4020, Epson MX/FX80, IBM Proprinter, Jlaser, HP Laserjet Plus, Apple Laserwriter
Typesetters Supported:	Linotype

DTP PACKAGES

Supplier Name and Address: Mirrorsoft Ltd
Headway House
66-73 Shoe Lane
London EC4P 4AB

Telephone Number: 01 377 4645

Product Name: Fleet Street Editor, Editor Plus, Publisher

Approx Retail Price: Fleet Street Editor:

BBC Micro	£39.95
Master Compact	£44.95
Amstrad CPC	£39.95
IBM PC V.2.0	£109.95

Fleet Street Editor Plus:

Amstrad PCW	£49.95

Fleet Street Publisher V.1.1:

Atari St	£125.00
PC TBA (Summer '88)	

Hardware Supported: —

Operating Environment: —

Page or Document Description Language Supported: Editor PC — PostScript, Publisher PC and ST, PostScript and Cora

Printers Supported: PostScript, HP Laserjet, Atari SLM 804

Typesetters Supported: Linotron 100/300/500 with PostScript RIP, Linotron 202 with Cora

Supplier Name and Address:	Prefis Ltd 64 Baldock Street Ware Hertfordshire SG12 9DT
Telephone Number:	0920 5890
Product Name:	Book Machine
Approx Retail Price:	£695-£1,900
Hardware Supported:	IBM, P.C.A.T. Compatibles, Apricot
Operating Environment:	MS-DOS
Page or Document Description Language Supported:	Cora
Printers Supported:	IPL Cora, Canon, Hewlett-Packard, Dot Matrix, NEC Epson, etc.
Typesetters Supported:	LINOTRON 100, 101, 202, etc.

DTP PACKAGES

Supplier Name and Address: Thames Valley Systems Ltd
Greys House
7 Greyfriars Road
Reading RG1 1NV

Telephone Number: 0734 581829

Product Name: Megascreen Macintosh System

Approx Retail Price: From £4,000

Hardware Supported: Macintosh

Operating Environment: Macintosh

Page or Document Description Language Supported: PostScript

Printers Supported: Any PostScript Laser and Apple Dot Matrix

Typesetters Supported: Linotype, Monotype, AM Varityper

Supplier Name and Address: Unifield Technology Ltd
8 Canal Street
Manchester M1 3HE

Telephone Number: 061 236 8406/0884

Product Name: TypeCast and TypeCast ATF

Approx Retail Price: TypeCast £249, ATF £510

Hardware Supported: IBM PC/XT/AT

Operating Environment: MS-DOS/PC-DOS

Page or Document Description Language Supported: PostScript and many native printer languages

Printer Supported: 9 and 24 pin DMP, most laser printers

Typesetters Supported: Linotronic via PostScript, Compugraphic MCS

DTP PACKAGES 211

Supplier Name and Address: VHA Computer Services
Howson-Algraphy Group
Vickers plc
Coal Road
Leeds LS14 2AL

Telephone Number: 0532 732442

Product Name: Document Management System (DMS)

Approx Retail Price: Complete system, micro-computer, laser and software £19,000, software only £7,950

Hardware Supported: Apricot Xen i 386/50 + 100 (Tower) IBM PC AT (X)

Operating Environment: DOS

Page or Document Description Language Supported: PostScript

Printers Supported: Various laser printers

Typesetters Supported: Linotronic 100, 101, 202, 300
Autologic APS5
Compugraphic 8000, 8400, 8600
MCS to Compugraphic 8600/8400/8000
Varityper Comp/Edit 5810 with digital ruling
Varityper Comp/Edit 6400/6410
Varityper Comp/Edit 6400/6410 with digital ruling
Varityper 6700/6730/6750
Atex front end
QMS-PS (with PostScript)
Tegra-Cora

Appendix F
Interfacing Typesetters

This is a method of obtaining typeset artwork from a phototypesetter that is remote from your own workplace by using the text already keyed into your own word processor. This is marked up for typesetting and gives you a degree of control over the layout.

There are now an increasing number of phototypesetting bureaux that offer the facility of interfacing through PCs. This can be achieved by a number of means. Text that is keyed in on your word processor can be interfaced with the typesetter by one of four major methods — telecommunications, text retrieval terminals, media convertors or direct disk reading.

The telecommunications option allows text to be transferred directly from your disk to a disk at the interfacing typesetter through a modem and telephone line. The advantage of this method is the speed with which text can be down-loaded to the interfacing typesetter and may be important if you need work processed quickly.

Text retrieval terminals are devices that plug in to your word processor via a standard RS232, V24 printer or communications port. They 'milk' the text from your disk and store it on a cassette which you send to the interfacing typesetter. In some cases, the typesetter can, where convenient, bring the TRT to you, down-load the data and then take it away for processing.

Media conversion and direct disk reading both involve sending the disk that contains the text files of the document to the interfacing typesetter to be typeset. If the disk is compatible with the interfacing typesetter's own equipment, then the text can be read into their machine without further processing. If the disk is not compatible with the hardware at the front-end of the typesetter, then it may have to be processed through a media convertor, which quite simply produces a disk that *is* compatible with the front-end of the typesetter.

Media conversion usually involves an extra charge on the job, however.

BENEFITS OF INTERFACING

If you have considered the typesetting of your technical documentation by conventional means to be expensive, you may find that interfacing using your WP is a cost effective way of getting typeset quality output, without having to invest in phototypesetting hardware yourself.

Interfacing typesetting involves you providing some coded version of your text by inserting commands which describe the format that the typeset output is to adopt. In this way you are reducing or eliminating (depending upon how much you do) the need for a photocompositor to insert phototypesetting commands or key in any original text.

Conventionally, manuscripts would be rekeyed by a photocompositor and coded up for layout, font styles, etc, and artwork sent back to the originator of the manuscript for checking. This method is still used in most book publishing and it is quite an expensive and labour-intensive way of typesetting.

One step further towards cost saving is to supply the manuscript on disk and thereby remove the need for rekeying. This has the added advantage of reducing rekeying errors to zero. However, you must ensure the accuracy of the text on the disk. The typesetter simply adds the commands in the text for phototypesetting and the job is processed at a reduced cost.

Interfacing typesetting goes just one step further. Not only does the author supply the text on disk but marks up the text with instructions about the style of the output (using what is known as 'generic coding').

INTERFACING TYPESETTERS

The extent to which these codes determine the final layout is dependent upon two factors — the degree of flexibility offered by the interfacing typesetter, and the amount of control you yourself want to exercise over the job in the first place.

HOW THE CODING WORKS

There are many variations of codes that can be used for typesetting by interfacing, and they may vary from one bureau to another. The 'standard' coding structure, as adopted by the print industry, is the one developed by Tony Randall, known as ASPIC. This is the acronym for Authors' Standard Pre-press Interfacing Codes, and is designed for people who have virtually no knowledge of typesetting terminology, or for people who do not know beforehand what the typographical parameters of the final job are going to be.

In essence, the coding structure is based upon a system of hierarchical codes that determine various levels of heading and text. For example, a level 1 heading (perhaps one that opens a chapter or section in the document) is assigned the code [h1]. The code is enclosed in some unusual character that is not often used in the text of the document, in this case, square brackets. What h1 translates to is determined by the interfacing typesetter and yourself, for example, it may mean a heading in 18pt Helvetica bold, centred with a 20pt advance leading and a 26 pica rule beneath. For simplicity's sake, a level 2 heading would be specified in your WP text as being preceded with an [h2] code, and this may define a whole new set of typesetting parameters.

Once the translation is agreed, the interfacing typesetter builds a translation database through which your text will be converted before being processed in the phototypesetter. In a similar way, all other format characteristics are determined and a tailor-made translation table is built for processing all jobs of a particular style and format for you. If you like, it is a kind of house style generator.

This method of coding is simple to use and can be adopted by anyone using a WP. All jobs can be coded in the same way, only the translation table determines what conversion takes place and therefore what the output will look like. You could supply to your WP operator, for example, the following list:

First level headings — chapter headings　　　　　　　　　　[h1]
Second level headings — sub-headings　　　　　　　　　　　[h2]
Third level headings — unnumbered headings　　　　　　　　[h3]
Left-hand page running heads　　　　　　　　　　　　　　　[h12]
Right-hand page running heads　　　　　　　　　　　　　　 [h11]
First level text — for main paragraph　　　　　　　　　　　 [t1]
Second level text — for paragraph numbers in left margin　　[t2]
Third level text — indented text　　　　　　　　　　　　　　[t3]

and so on. In addition, other codes are used to make changes in the text for style changes, for example [i] for italics, [b] for bold, [r] for roman (return to normal).

All documents could be coded in this way by the person responsible for keying in text, and can be done either as the text is first keyed or as a separate exercise to be performed once all the text has been entered.

A translation table is then agreed which determines what the codes will translate to. For example:

[h1]　This code converts into 14 point Helvetica bold
[h2]　This code converts to 11 point Helvetica bold
[h3]　This code converts to 11 point Helvetica bold indented 2 picas
[h11]　This code converts to 9 point Helvetica, ranged right
[h12]　This code converts to 9 point Helvetica ranged left
[t1]　This code converts to 11 or 12 point Helvetica justified over a measure of 24 picas, indented 2 picas

and so on.

The power of such a system is that, from the user's point of view, the coding need not get any more complicated. From the interfacing typesetter's point of view, however, a simple code could be converted into a whole string of typesetting parameters which perform a whole range of formatting commands affecting page style, point size, font choice, kerning, character compensation, hyphenation requirements, justification, measure, etc.

ASPIC has been adopted as the industry recommended standard by the British Printing Industry Federation, and you may well

INTERFACING TYPESETTERS

encounter ASPIC being offered by typesetters as their interfacing code system. Variations on ASPIC also exist, but they are nearly always based on similar principles.

GETTING MORE CONTROL

Coding systems are available to give even more control over the typesetting parameters than ASPIC. The main advantage is that no individual translation is required for each type of specification. This can reduce initial costs since the cost of setting up an ASPIC translation table has to be met by the typesetter's client. Any change to the original specification requires another translation table and more cost. If your work is quite variable, you may well prefer to use a more flexible coding system. A disadvantage of this is the additional keystrokes required due to the more complex level of coding.

One of these coding systems is MAGIC, which stands for Manager's Alphanumeric General Interfacing Codes. Also the work of Tony Randall, this is a print-orientated set of codes. Like ASPIC, MAGIC codes are embedded into the text of the document to specify the typesetting parameters. The standard set of codes is compatible with the Compugraphic 8600 digital phototypesetter driven by an MCS front-end, but is capable of being adapted to any other system with similar characteristics.

The difference with MAGIC, compared to ASPIC, is that each code converts into one particular typesetting command. For example, to specify a measure of 18 picas, font library number 24 (Garamond), point size 14 point, and leading of 18 points the following codes would have to be entered in the WP text:

[m18] [f24] [s14] [d18]

Such a coding structure is much closer to the user entering the typesetting commands directly, and consequently offers the benefits of more flexibility than ASPIC.

WHERE TO GO FOR INTERFACING TYPESETTING

For the sake of convenience it is worth checking whether your local facilities for typesetting can offer you an interfacing service for your typesetting requirements. If they do, you will need to establish how they go about transferring your data to their equipment and what costs are involved for this, particularly if your systems are incompatible and media conversion is required.

Check also how much interfacing they have done. If they are inexperienced in this field, you may become the guinea pigs for solving the difficulties faced.

If you have any difficulty in finding a reliable interfacing typesetter, then I recommend you go to the people who invented ASPIC, who are more than willing to help newcomers to this field get to grips with this method of typesetting. They are also fast and cost effective when compared to any conventional means.

Contact: Tony Randall
Electronic Village Limited
1 Orchard Road
Richmond
Surrey TW9 4PD

Telephone: 01 876 7013

SYNOPSIS OF ASPIC

Headings
[h1] Starts 1st level
[1x] Ends 1st level
[h2] Starts 2nd level,&c
[2x] Ends 2nd level, &c

Text
[t1] Starts 1st level
[t2] Starts 2nd level

Indented text sections
[i1] Starts 1st level indent
[i2] Starts 2nd level, &c
[ix] Ends indent

Font changes
[r]   Roman, or [ro]
[i]   Italic, or [it]
[b]   Bold, or [bo]
[sc] Small caps
[sx] End small caps

Justification
[jc] Centred
[jj] Justified
[jl] Range left
[jr] Range right

Ending text paras
]]   Flush left, new line
[]   As above, plus indent
[[New line

INTERFACING TYPESETTERS

Gaps for illustrations
[qp] Space for quarter page
[hp] Space for half page
[wp] Space for whole page

Specialised codes
[id] Insert leader dots
[ir] Insert baseline rule
[is] Insert space
[iv] Insert vertical space
[lc] Line centre
[ll] Line left
[lr] Line right

Supplementary ASPIC
[ad] Arrow down
[al] Arrow left
[ar] Arrow right
[au] Arrow up
[at] At sign @
[bl] Brace left {
[br] Brace right }
[bu] Bullet ●
[co] Copyright sign ©
[lt] Less than sign ⟨
[mt] More than sign ⟩

Appendix G
Organisations

The following pages contain some address details of organisations that may be of help to support the working technical author or publications manager. These details, however, such as telephone numbers etc, may be subject to change without prior notice from the publishers.

PROFESSIONAL BODIES FOR AUTHORS:

The Institute of Scientific and Technical Communicators
52 Odencroft Road
Britwell
Slough
Berkshire SL2 2BP

Telephone: (0753) 691562

The ISTC is a professional organisation for those involved in the communication of scientific and technical information. Their objectives are to establish and maintain professional standards, to assist in and develop training, and to provide a source of information to members. The ISTC publishes a quarterly journal called *The Communicator* with a supplementary newsletter published six times a year.

**The Society of Authors
84 Drayton Gardens
London SW10 9SB**

Telephone: 01-373 6642

The society, founded in 1884, is an independent trade union which represents, assists and protects a membership of almost 4,000 authors. The permanent staff have wide professional experience of the many problems encountered by authors in general, and have immediate access to lawyers, accountants and insurance consultants, retained by the society.

While the society deals with the authors' requirements in the field of general publishing, they also have specialist groups, which include the Science, Technical and Specialist Group who publish their own bulletin.

Science, Technology and Specialist Group:
This is a specialist group within the Society of Authors. The group's activities are reported in the Noticeboard section of the SoA's publication *The Author*, and publish their own official newsletter called *The Stag*. See address details for the Society of Authors.

PHOENIX SEMINAR FOR PUBLICATIONS MANAGERS

Publications managers and authors tend to work in isolated groups and they rarely have the chance to meet their opposite numbers from other companies.

The Phoenix seminars for publications managers provide a forum for informal discussion and organised talks on a variety of topics associated with the requirements and problems of technical documentation production. Currently held twice a year, the seminars are administered and organised by Phoenix Technical Publications Limited who act as secretariat.

In order for future seminars to be assured, and to increase the communication of ideas and information in the field of technical publications, Phoenix are anxious to encourage more attendees to the seminars. If you would like further details, please contact:

ORGANISATIONS

John Gardner, Director
Phoenix Technical Publications Ltd
Barford House, Shute End
Wokingham
Berkshire RG11 1BJ
Telephone: (0734) 774211

TRANSLATION SERVICES

The Institute of Translation and Interpreting
318a Finchley Road
London NW3 5HT

Telephone: 01-794 9931
Fax: 01-435 2105

The Institute is a professional association for translators and interpreters. They can be a source of information about translation services and can provide details of members capable of handling particular language and subject combinations.

**Interlingua TTI
Imperial House
15-19 Kingsway
London WC2B 6UU**

Telephone: 01 240 5361
Telex: 95101 Lingua G
Fax: 01 240 5364

Incorporating LinguaSoft — a specialist software translation company.

Interlingua TTI is part of an international network of translation companies which offers a complete service from advice on adapting a product to a foreign market, to translation of all types of technical documentation and software programs, through artwork, design and typesetting, to print. Interlingua TTI translates 85 languages, and has particular expertise in the computer, defence, automotive, pharmaceutical and telecommunications industries.

Part of the a.l.p. Systems network, Interlingua TTI uses a.l.p. Systems computer aided translation tools for certain projects, ensuring efficiency, ability to interface with clients' source and target production requirements, and high quality consistent text.

Interlingua TTI has offices in London, Bristol, Birmingham, Nottingham, Leeds, Newcastle, Manchester, Hong Kong, Singapore, Barcelona and Madrid, with network offices in the USA, Canada, France, Germany and Switzerland.

Hook and Hatton Ltd
34 Central Avenue
Whitehills
Northampton NN2 8DZ

Telephone: 0604 847278
Fax: 0604 847278 & 24282
Telex: 94011648 HOOK G

Hook and Hatton Ltd specialise in the preparation, translation and publication of technical manuals. Their services include technical, legal and commercial translations; technical writing; editorial assistance; copywriting; word processing and mailing; foreign language artwork and printing.

SUNDRY ADDRESSES

The Society of Indexers
16 Green Road
Birchington
Kent CT7 9JZ

Telephone: (0843) 41115

This organisation is a professional body that looks after the interests of indexers in terms of improving standards and salaries, and advising on qualifications and remunerations. The society publishes and communicates through books, papers, journals, etc on the subject, and maintains a register of indexes.

Screen Dump Software Utility (For IBM PC/XT/AT)

Bates Associates
64 Welford Road
Wigston Magna
Leicester LE9 1SL

Telephone: (0533) 883490

ORGANISATIONS

Bates Associates provide three resident screen dump programs, written in 8086/8088 Assembler and once loaded (at the DOS prompt) they remain resident until the machine is rebooted. In each case, the resident code replaces the existing INT 5 PrtSc routine. This means that the operator simply presses the Shifted PrtSc key to suspend current program operation and print an exact copy of the screen. The printout is produced sideways and is easily accommodated on 80 column printers. Once the printout is completed, program operation resumes as normal. Versions for EPSON and IBM dot matrix printers are available.

Appendix H
Further Reading

Austin Mike, *Technical Writing and Publication Techniques*, ISTC/William Heinemann, 1987

Bly R W, Blake G, *Technical Writing — Structure, Standards and Style*, McGraw-Hill, 1982

Brockman R John, *Writing Better Computer User Documentation — from Paper to Online*, John Wiley & Sons, 1986

Browning C, *Guide to Effective Software Technical Writing*, Prentice Hall, 1984

Evans J, *Beginner's Guide to Technical Writing*, Newnes Technical Books, 1983

Fowler H W, *A Dictionary of Modern English Usage*, Oxford University Press, 1968

Grimm S J, *How to Write Computer Manuals for Users*, Wadsworth, 1982

Hartley J, *Designing Instructional Text*, Kogan Page, 1985

Jones Graham, *The Desktop Publishing Companion*, Sigma Press, 1988

Kelly Derek, *Documenting Computer Application Systems*, Petrocelli Books, 1983

Lang Kathy, *The Writer's Guide to Desktop Publishing*, Academic Press, 1987

Lomax J D, *Documentation of Software Products*, NCC Publications, 1977

McKay Lucia, *Soft Words Hard Words — A Commonsense Guide to Creative Documentation*, Ashton-Tate, 1984

Miles John, *Design for Desktop Publishing*, Gordon Fraser, 1988

Price Jonathan, *How to Write a Computer Manual — A Handbook of Software Documentation*, Benjamin Cummings, 1984

Turk C, Kirkman J, *Effective Writing*, E&F Spon, 1982

Worlock Peter, *The Desktop Publishing Book*, Heinemann, 1988

Zaneski R, *Software Manual Production Simplified*, Petrocelli Books, 1982

Appendix I
Glossary

The following are some of the terms associated with desktop publishing and typography that you may find useful.

Baseline

Different typefaces vary in size and style and, so that they can be mixed on the same line, are aligned on an imaginary horizontal reference line known as the baseline.

H H H H

Bit Mapping

This relates to the building up of an image using a matrix of dots. Scanners and laser printers use bit mapping for processing graphics. The quality of the image produced is dependent upon the resolution or density of dots in a given area. This is usually expressed as a number of dots per inch, for example 300dpi.

Bromide

This is the photosensitive paper on which the output of a phototypesetter in developed through a processor, not unlike a film processing system. In fact, the term 'film setting' is still used by some people, and the bromides are referred to as the film output.

Condensed Type

This is not often available on DTP systems and is found as a facility of some digital phototypesetters. The term itself refers to the relative narrowness of characters in a particular typeface. For a digital typesetter, changing what is known as the 'set width' can invoke the condensing of the characters. For example, a 14 point character can be defined as having a set width of 12 points, which reduces the size of the character's width rather than simply adjusting the space between the characters.

This is 14pt with normal set width
This is 14pt with a set width of 12pt

Expanded Type

This is effectively the opposite to condensing type. For digital typesetters, altering the set width can cause the characters of a particular typeface to be expanded horizontally so as to occupy more width, for example, a 12 point character with a set width of 14 points.

This is 12pt with normal set width
This is 12pt with a set width of 14pt

Fixed Spaces

In typography, where certain fixed spaces between characters are required, a system of spacing based on point of space is employed to adjust the space between words or characters that is otherwise variable, depending upon the justification of the text on the line. There are three commonly used spacings:

Em Space — generally, this is a space which is equivalent to the value of the point size, so that a 14 point em space will be 14 points wide.

En Space — this is half an em space.

Thin Space — this may be either ¼ or ⅓ of an em space.

These spaces are used to make adjustments where additional fixed space is required, for example the first line of most of the paragraphs within this book are indented from the left hand margin.

Fount (Font)

A complete set of characters in the same typeface and size, including letters, punctuation and symbols. For example, 12pt Garamond is a different fount to 12pt Garamond italic, or 12pt Univers, etc.

Galleys

The word galley is generally only used in phototypesetting since it refers to a continuous length of typeset material which may be used for proofing or cutting and pasting into position according to the layout requirements. Whereas DTP systems are designed to control the format and layout of the text before output, commercial typesetters would often provide the setting as lengths of text, set at the appropriate width. If this was then required for a manual, for example, the galleys would need to be cut into page size, in order to achieve the pagination.

The term originates from the 'hot metal' process of typesetting, where a metal tray with raised edges was used to hold about 20 inches of metal type. In this context the term galley referred to the amount of text set, but the word is more generally used to define the status of a job, ie a job which has reached 'galley stage'. For example, in book publishing it is usual for a book to be set in galleys, then these are proofed back to the author for checking. Once the final setting alterations are made, the galley is turned into paginated or formatted output according to the design requirements of the job.

Headers and Footers

Also known as running headers or footers. Footers are also referred to as trailers. These are one or more lines of text that appear at the top or bottom of a document's page for including detail such as chapter headings, reference numbers, folios, etc.

Hyphenation

This is the process of deciding where to hyphenate the last word on a line. Some systems will do this automatically, based on a hyphenation dictionary, but you may wish to override this and include your own.

Indents

Space defined at either end of the line to change the placement of the text. The most common form of indent is for a paragraph denoting the beginning of a block of text. Generally, you should make the indent space proportional to the length of the line, for example:

lines under 24 picas wide — indent 1 em space
lines between 25-36 picas wide — indent 1½ em spaces
lines 37 picas wide or over — indent 2 em spaces

Justified Text

Text that is aligned on both the left and right hand margins of the document, as used in this book.

Kerning

Kerning is the reduction of space between characters of a typeface in order to improve the appearance when certain pairs of characters appear together. For example, the letter W, when typeset, has a certain amount of optical space beneath the slant of the right and left downstrokes, and this can vary from typeface to typeface. If the letter following it, for example, is set with the normal letterspacing, it can appear to be displaced too far to the right. By subtracting points of space from between the characters, for example, the following letter effectively overlaps into the set width of the W, making the appropriate optical correction.

WA	normal letterspacing
WA	minus 1 point of space
WA	minus 2 points of space
WA	minus 3 points of space
WA	minus 4 points of space

    Paired kerning is the term used where a composition software system identifies the pairs of characters that require space adjustment and applies kerning automatically, provided the facility is switched on. Generally, the kerning requirement is more noticeable the larger the size of the typeface.

GLOSSARY

Leading

The amount of vertical space, expressed in points, between the baselines of two lines of text (it is pronounced 'ledding').

Letterspacing

This is the term that refers to the amount of space between individual letters, and on most phototypesetters is adjusted by subtracting or adding points of space. This is known as negative or positive letterspacing respectively. Positive letterspacing may be used to put space between the letters of words, rather than have larger gaps between the words themselves which might make them look spread along the line width. Sometimes, positive letterspacing may be used to achieve a particular effect, for example, to space out the characters of a heading.

Generally, positive letterspacing interrupts the legibility of the text, so negative letterspacing is the more commonly applied adjustment. This is used to tighten the spacing between letters to improve the appearance of the text, adjust space for *kerning* or to fit copy into a tight space.

Page Description Language (PDL)

This is a software facility which is independent of the hardware of desktop publishing systems. It is used to convert a screen image which may include text and graphics, into instructions that can be used to drive an output device like a laser printer. Using such an independent interface allows control over text and graphics, and multiple typestyles and sizes. Postscript is probably one of the best known page description languages, and is generally regarded as the current standard. It has been adopted by a wide range of manufacturers, including IBM and Apple, and offers a path to phototypesetting from PC front ends. Postscript was developed by Adobe. Other PDLs include Xerox Interpress, Imagen Impress and Imagen/HP DDL.

Page Make-up Software

This is software that can produce 'compound documents' which can comprise of text and graphics together. Using the concept of electronic paste-up, page make-up software can use text files generated on a word processor for pasting into a page layout. Graphics can be similarly imported or scanned in directly. Page make-up software normally uses a page description language. The distinction of such a package is that text-only software, such as a word processor, has limited output capabilities without a PDL, and provides limited control over layout and typefaces.

Pagination

Pagination is the process of assembling text, graphics, running headers, folios, etc, to produce pages of the required length ready for printing. This can be done by cutting and pasting-up setting and illustrations etc by hand, or electronically using a DTP or phototypesetting system.

Picas

This is a unit of measurement equal to $1/6$ of an inch (see also *point system*).

Point System

A point is a unit of measurement generally considered equal to $1/72$ of an inch. However, there have been three point systems introduced:

The American-British system — point is measured as 0.1383 inch, or one twelfth of a pica (pica being .166 inch).

The Didot system — basic unit is the cicero, which is equal to 12 corps (points) or .178 inch. The Didot corps measures exactly .01483 inch.

The Mediaan System — point (or corps) measures .01374 inch.

For general purposes, it is useful to remember: 6 picas to one inch; 72 points to one inch; 12 points to one pica, though these are not strictly accurate since the point does not directly relate to inches.

Pi Characters/Founts

These are special characters or symbols such as ⅛, ¼, ⅜, +, @, 〈, 〉, etc. These may be used for maths requirements in text, or simply decorative purposes.

Quadding

A term relating to the placement of text. The word originates from 'quadrat', which was a metal cube used for filling blank space in hand typesetting. Quads, therefore, are used to specify where the remaining space on a line is to be positioned.

This is quad left

This is quad right

This is quad centre

Raster Image Processor

This device is the basis of truly controllable desktop publishing. It converts instructions into a bit map (see *bit mapping*) and allows a laser or page printer to use a page description language (see *page description language*).

Set Width

The set width of a character relates to its point size, so a 9pt character has a set width of 9. Some typefaces, however, are designed with a narrower set width, eg 9pt with 8.5 set width.

Small Caps

These are capital letters designed to match the x-height of a typeface, though many founts these days do not have small caps so they are created by reducing the point size to 80% of its original size. They are used for abbreviations of awards, titles, etc following a name, or where whole words in text are set in all caps.

CAPS and SMALL CAPS

Superior/Inferior Characters

These terms relate to characters usually set in a smaller size to the text typeface, and positioned above (superior) or below (inferior) the baseline. They are often referred to in DTP systems as superscripts and subscripts respectively.

$A^2 \quad B_4$

Type Size

Type sizes are described by the basic unit of measurement, the point. Points are used to define the length of a metal block or chunk of type. The length of the metal block's top surface relates to the point or type size, though the character cast onto the surface will be smaller than the overall size of the metal. Traditionally, then, the point size of a typeface refers to the dimension of the metal block, not the height of the image. This is to allow room for the ascenders and descenders. Thus the type size relates to the distance between the top of the ascender and the bottom of the descender of the typeface.

Weight

The weight of a typeface varies according to its design. The thickness of line will determine how light or dark its image will appear after printing. There are standard terms for the various weights of a typeface and these are referred to as extralight, light, semilight, regular, medium, semibold, bold, extrabold and ultrabold. However, these variations are not standard across all typefaces. For example, 'medium' weight of one typeface may be the same weight as 'bold' in another.

Word Spacing

In phototyesetting the space between words is variable (unlike a typewriter), which enables lines of text to be justified. It also allows for the optimum number of words to be set on the specified line length.

WYSIWYG

(W)hat (Y)ou (S)ee (I)s (W)hat (Y)ou (G)et. This term is used to describe a feature of word processors or desktop publishing software systems that can represent on the screen display what the actual output of the document will look like. The degree of accuracy in this respect, however, varies greatly from system to system.

x-Height

This is the height of the letter x and is used to describe the height of the body of lower case characters in a particular typeface.

GLOSSARY

Ascender
x-height
Descender

A change in x-height can effect the apparent size of typefaces that are actually the same size.

This has a tall x-height — set in 12pt Benguiat
This has a low x-height — set in 12pt Palacio

Index

Abbreviations — house style	42
Ambiguities in documentation	40
Amendments — scheduling for	155
Apple Macintosh	108, 110
Artificial intelligence — in DTP packages	118
Artwork	50
Artwork preparation — scheduling for	156
Ascenders — typefaces	68
ASPIC	Appendix F
Attention to detail	16
Authoring first draft	155
Author recruitment	Appendix B
Author's role	11-22
Author training	Appendix C
Automatic indexing	98-99
Baseline	Appendix I
Bibliography	Appendix H
Binding	171
Bit image	76
Bit mapping	78, App I
Body size — typefaces	65, 68
Books — for authors	Appendix H
Bromide	Appendix I
Budgets	142-143
Building a schedule	151-152

CAD/CAM	76, 80
Capitals — house style	41
Chapter beginnings and headings	61
Chromalin proofs	168
Colour	
— presentation of documentation	169-170
— use of	73
Colour printing — considerations	168
Columns — multiple, layouts	48, 57
Commercial awareness — author	20
Communicating with internal sources	29
Communication skills — of author	16
Competition — assessing the	31
Completeness of documentation	37-39
Computer-aided text processing	107
Condensed type	Appendix I
Consistency of documentation	40
Contract authorship	25, 26, 28
Contractors	Appendix A
Copy editing and proof correction	Appendix D
Copyright and trademarks — house style	43
Corporate electronic publishing	107
Costing and scheduling	138-139, 141-149
Costing resource time	143-146
Cost per page	147-148
Daisywheel printer	93
Dates — house style	42
Descenders — typefaces	68
Design and layout	47-73
Design concept — general	152, 155
Desktop publishing	107-119
Development — documentation	183-185
Didot, Firmin	65
Digitex — national survey	24
Documentation	
— evaluaton of	131-133
— important factors of	14-15
— planning of	128-129

INDEX

Documentation houses	25, 26
Dot matrix printers	93
Drafting text on word processors	103, 105
Drawing software	78
Drawings	80-81
DTP	
— identifying your needs	111-112
— making good use of	113-114
DTP packages — supplies of	Appendix E
DTP users	110
— cautions for	116-119
Economic control — author	18
Editing — scheduling for	156
Electronic pre-press publishing	107
Electronic proof readers	99-103
Electronic scheduling systems	159
English — general use of	33-36
English skills	14
Evaluating documentation	131-133
Expanded type	Appendix I
Expert systems — DTP	118
External costs	147
Figures — house style	42
Finished artwork	50
First draft — authoring of	155
Fixed spaces	Appendix I
Font (fount)	Appendix I
Footnotes and running heads	61, 65
Format and style	33-45
— points to consider	36-40
Format — presentation of documentation	169
Freelance authors	22, 28
Further reading	Appendix H
Galleys	Appendix I
Glossary	Appendix I
Graphics/drawing software	78-80

Gravure printing	163, 166
Grey scales — scanners	78
Gutenberg, Johann	65, 161
Halftones	81, 83
Headers and footers	Appendix I
Headings	
— chapters	61
— house style	41
Help text	179
Highlighting text — house style	42
House styles	40-43
Hyphenation	Appendix I
— house style	41
Illustrated concepts	43
Illustration	75-89
Illustrations	
— presentation of documentation	170
— scheduling for	155-156
— use of	51
Indents	Appendix I
Indexing	
— house style	43
— with word processors	98-99
Inferior characters	Appendix I
In-house production — justification	138
Ink-jet printers	93
Intaglio	166
Interfacing typesetters	Appendix F
Internal sources — communicating with	29
ISTC (Institute of Scientific and Technical Communicators)	26
Jargon	34, 35
Justified text	Appendix I
Kerning	Appendix I

INDEX

Laser printers — how they work	114
Laserwriter	108
— with word processors	93, 96
Layout and design	47-73
Leading	Appendix I
Leads — typography	68
Letterpress	162-163
Letterspacing	Appendix I
Line copy/drawings	80-81
Line diagram	75
Line/halftone combinations	83
Linotype	162
Lists — house style	42
Lithography	161, 166-168
MacAuthor	108
MacDraw	108
MacWrite	108
Management reporting — author	18
Management skills — author	18
Maps — house style	42
Material cost sheets	146-147
Methods of printing — paper use	170-171
Monotype	162
Multiple columns	48, 57
Near letter quality print	93
Nick — on type chunk	68
Notes and reference systems — house style	42
Offset printing	163, 166-168
On-line documentation	179-181
Organisations — sundry	Appendix G
Outside writing sources	24
Overhead cost sheets	146-147
Ozalid proofs	167
Page description language (PDL)	Appendix I
Page Maker	108

Page make-up software	Appendix I
Page style — house style	41
Pagination	Appendix I
Paper — quality of	170-171
Paragraph spacing	61
Photoengravure	163
Phototypesetters — how they work	124-126
Phototypesetting	121-126
— by interfacing	Appendix F
— via word processors	96
— why use?	123-124
Pi characters/fonts	Appendix I
Picas	68, Appendix I
Planning of documentation	128-129
Planning type area on page	56-61
Plates — house style	42
Points	65, 68, Appendix I
Presentation and packaging	169-172
Print buying	133-138
Printers	93
Printers and printing	161-168
Printing and finishing — scheduling for	156, 159
Progress chasing	139
Project management	26
Proof correction	Appendix D
Proofing cycles	26, 29, 32
Proofing — scheduling for	155
Proof readers	
— word processors	101-103
— electronic	99-103
Publications manager — role of	127-140
Punctuation — house style	41
Qualities — of author	17-18
Quadding	Appendix I
Raster image processor	Appendix I
Reader knowledge	13
Reading — list of	Appendix H

INDEX

Ready Set Go	108
Recruiting of authors	129-131, Appendix B
Reference systems — house style	42
Relevance of documentation	39-40
Resolution — laser printers	116
Retouching	83, 85
Review of documentation	44
Revisions and updates	173-178
Running headers	61, 65
Scanners	76-78
Scheduling	138-139, 151-159
— electronic systems for	159
Screen illustrations	
— software documentation	85, 88, Appendix G
Screening	81, 83
Set width	Appendix I
SI units — house style	42
Small caps	Appendix I
Software documentation	
— screen illustrations	85, 88, Appendix G
Solid setting	68
Spaces	Appendix I
Spelling	15
Spelling and hyphenation — house style	41
Spelling checkers — on word processors	99-100
Stock control	139-140
Structure of text	43
Subject knowledge	12
Superior characters	Appendix I
Superpage	108
Survey — of computer documentation	24
Tables — house style	42
Thesauruses — on word processors	100-101
Time management — author	18
Time sheets	146
Trademarks — house style	43
Training for authors	Appendix C

Training notes	48
Translation services	Appendix G
Type	
— fundamentals of	65
— metal chunk of	65
Type area — planning on page	56-61
Typefaces	65, 68
Type size	Appendix I
Typing — scheduling for	155
Units — house style	42
Updating documentation	173-178
Web-fed processes	166
Weight	Appendix I
White space — on page layouts	56
Word spacing	Appendix I
Word processing	91-105
— in schedules	155
Word processor layouts/designs	51
Word processors	
— drafting text	103, 105
— indexing with	98-99
— layout with	96, 98
— types	92
Writing sources	23-32
— outside	24-29
WYSIWYG	92, 93, Appendix I
X-height	Appendix I